NAVAL FIREPOWER

A US Battle Squadron is shown during gunnery practice in 1921. The leading ship is *Nevada* or *Oklahoma*, distinguishable by her pattern of secondary gun mounts. Note the characteristic US cage masts, intended to insure that ships could keep firing at long range despite mast hits, and the bearing markings on No 2 turret.

NAVAL FIREPOWER
BATTLESHIP GUNS AND GUNNERY IN THE DREADNOUGHT ERA

NORMAN FRIEDMAN

Drawings by A D Baker III and W J Jurens

Seaforth
PUBLISHING

Copyright © Norman Friedman 2008

First published in Great Britain in 2008 by
Seaforth Publishing
An imprint of Pen & Sword Books Ltd
47 Church Street, Barnsley
S Yorkshire S70 2AS

Website: www.seaforthpublishing.com
Email: info@seaforthpublishing.com

Reprinted 2008

British Library Cataloguing in Publication Data

Friedman, Norman, 1946-
Naval Firepower : battleship guns and gunnery in the dreadnought era
1. Fire control (Naval gunnery) - History 2. Battleships -
History 3. Naval art and science - History - 19th century
4. Naval art and science - History - 20th century 5. Naval
history, Modern - 19th century 6. Naval history, Modern - 20th century
I. Title
623.5'53'09

ISBN 978-1-84415-701-3

All rights reserved. No part of this publication may be reproduced or transmitted in any form or by any means, electronic or mechanical, including photocopying, recording, or any information storage and retrieval system, without prior permission in writing of both the copyright owner and the above publisher.

The right of Norman Friedman to be identified as the author of this work has been asserted by him in accordance with the Copyright, Designs and Patents Act 1988.

Designed and typeset by Roger Daniels
Printed and bound in Thailand

Contents

	Note on units of measurement	6
	Note on abbreviations	6
	Author's acknowledgements	7
	Introduction	8
1	The Gunnery Problem	16
2	Range-keeping	40
3	Shooting and Hitting	66
4	Tactics 1904–14	82
5	The Surprises of War 1914–18	100
6	Between the Wars	112
7	The Second World War	140
8	The German Navy	156
9	The US Navy	156
10	The US Navy at War	206
11	The Imperial Japanese Navy	227
12	The French Navy	242
13	The Italian Navy	256
14	The Russian and Soviet Navies	268
	Appendix: Propellants, Guns, Shells and Armour	280
	Notes	292
	Glossary	313
	Bibliography	315
	Index	316

Note on units of measurement

I have used predominantly Imperial units of measurement (mainly yards and inches) rather than metric, because they were standard in the two most powerful battleship-era navies, the US Navy and the Royal Navy. Gun calibres in inches will be familiar to most readers. However, in those chapters about navies that used metric measurements, I have given the metric measurement first, with the Imperial conversion in brackets. Before the adoption of metric units, the typical unit for range was the yard; in gunnery 2000 yards was used for the nautical mile (it is actually slightly longer). A metre is slightly longer than a yard (1 metre is about 1.09 yards). For weights, a pound is slightly less than half a kilogram (1 kilogram is 2.204 pounds).

Notes on abbreviations

A glossary of terms and abbreviations is provided on page 313. In the main text, terms set in italics are defined in this glossary.

Because they recur frequently in the endnotes (pages 292–312) I have used the following abbreviations to indicate some sources:

DNOQ: Principal Questions Dealt with by DNO (1889–1911 in PRO) and Important Questions Dealt with by DNO (1912–14 in NHB).

DGBJ: Brooks, John, *Dreadnought Gunnery and the Battle of Jutland: The Question of Fire Control* (Abingdon: Routledge, 2005).

EHL: Historical Library at HMS *Excellent* at Portsmouth.

IDNS: Sumida, Jon Tetsuro, *In Defense of Naval Supremacy: Finance, Technology, and British Naval Policy 1889–1914* (Boston: Unwin Hyman, 1989).

NARA: US National Archives and Records Agency.

NARA II: US National Archives and Records Agency outpost at College Park, Maryland.

NHB: Naval Historical Branch at Portsmouth.

PRO: The National (British) Archives at Kew (formerly the Public Record Office).

SHM: The Service Historique de la Marine of the French defence archive at Vincennes.

Details of US systems were generally taken from the:

OPs: Ordnance Pamphlets, which are often actually thick bound books).

ODs: Ordnance Data, usually typescript equivalents of OPs produced by blueprint, at College Park, Maryland; the main exception is the Mk 31 director.

The Ordnance Specifications (**OS**) for specific classes in the pre-World War II Bureau of Construction and Repair (RG 19) files at the US National Archives in central Washington describe many ships' systems (they are much more detailed and specific than their title suggests).

Author's acknowledgements

Professor Jon Tetsuro Sumida, who should be credited with reviving the understanding of the importance of fire-control technology in twentieth-century naval warfare, was particularly generous with documents and advice. Dr Nicholas Lambert provided much valuable British material, and Stephen McLaughlin helped with extensive Russian material. Christopher C Wright, editor of *Warship International* and author of a series of recent articles on US battleship fire-control systems, shared his research material and also provided advance copies of some of his articles. Alexandre Sheldon-Dupleix provided material from the French archives, including a copy of the 1933 Italian naval fire-control text, which the French had obtained and translated (Dr Paul Halpern told me about this document). He also provided photographs and material from the trials reports of the battleship *Dunkerque*. Andrew J Smith provided Italian material. Steve Roberts provided some essential French material as well as invaluable advice on scanning. Mark Wertheimer of the US Naval Historical Center provided some important US documents.

For assistance with the US National Archives I am particularly grateful to Barry Zerby and Ken Johnson at College Park, Maryland and to Charles Johnson at the downtown archives. Captain Christopher Page RN (Ret'd) and his team at the Royal Naval Historical Branch (particularly Admiralty Librarian Jenny Wraight and Admiralty Archivist Kate Tildsley) in Portsmouth were extremely helpful, as was Lieutenant Commander Brian Witts RN (Ret'd), the curator (and creator) of the HMS *Excellent* historical collection. I am also very grateful to the staff of the Brass Foundry (an outpost of the National Maritime Museum) and to that of The (British) National Archive, which I continue to think of as the Public Record Office. John Spencer provided invaluable help navigating the French naval archives, as well as considerable other assistance. Randy Papadopulous of the US Navy Operational Archives helped with documents from the US Naval Technical Mission to Europe and with some US BuOrd material. The staff of the US Navy Department Library helped me find and copy some rare material including late interwar annual gunnery reports.

Most of the photographs were provided by Charles Haberlein (and his assistants Ed Finney and Robert Hanshew) of the photographic archives of the US Naval Historical Center and by A D Baker III, who also drew some of the illustrations. Dr Thomas C Hone provided some photographs, including the ones taken before the Bikini tests. John Asmussen of the *Bismarck* website (www.bismarck-class.dk) provided valuable photographs. I am grateful to William Jurens for many drawings, which are modified versions of ones he originally produced for *Warship International*. A D Baker III also created original drawings for this book. Except as noted at the end of individual captions, photographs are from the Naval Historical Center, the US National Archives and private collections.

The manuscript for this book was read in whole or in part (in various versions) by (in alphabetical order): A D Baker III, Christopher Carlson, Trent Hone, David C Isby, Dr Nicholas Lambert, Stephen McLaughlin, Andrew J Smith, John Spencer, Dr Jon Tetsuro Sumida, and Dr Alan Zimm. All provided helpful comments. The opinions expressed in this book are my own, however, and should not be attributed to any of my readers or helpers. Whatever errors have survived their perusal are entirely my own.

I could not have written this book without the advice and loving support of my wife Rhea. The research for this book would have been impossible without the use of an electronic camera and a scanner. Rhea encouraged me to use a camera for research many years ago, and also encouraged me to take the many and frequent trips to distant archives (London, Portsmouth, Paris, Washington) necessary to complete this project.

hit at longer ranges because they were so much steadier. Scott had carefully observed gunners, some of whom achieved great success by continuously elevating and depressing their guns so that they were always on target. Scott called this technique continuous aim. It was no longer necessary to choose a point in the ship's roll at which to fire. Waiting for the gun-sights to come 'onto' the target had always been a source of error, because no gunner's reflexes were instantaneous. Moreover, firing only at a set point in the roll limited the firing rate.

Scott was stabilising guns *in the line of sight* but not across it (ie against cross-roll). This pre-World War I gunnery revolution extended line-of-sight compensation to heavy guns so that, by 1914, gunners firing at 10,000 yards could make many more hits than they might have at 1000 yards before Scott. Cross-roll, however, was a different proposition.

Scott's technique changed the roles of those controlling the guns. In the past, *laying* the gun (elevating it) had meant setting it at a fixed elevation for the ordered range. Now the *gunlayer*, who kept the gun on target, had the key role. Because he could tell whether the gun was on target, he was the one who fired. Scott introduced a separate sight-setter to enter the required elevation. As before, *pointers* or *trainers* pointed the gun on the appropriate bearing (the US Navy called the pre-Scott technique *pointer firing* because pointers had been more important than layers).

Scott replaced the earlier method of trying to fire when the gun reached a set point in the ship's roll. The ship was momentarily at rest at the top and bottom of a roll. It moved fastest at mid-roll, but that speed (up or down) was more or less constant for a time. Although many gunners advocated firing at the top or bottom of a roll, it was not always easy to sense that this end of the roll was approaching, and it was very easy to miss such moments. Some ships made that easier, because they had a steady, predictable roll. Training helped ensure that gunners who sensed that the ship was approaching one end of the roll could move quickly enough for their guns to fire at the right moment (or at least close enough to it). However, if the ship was rolling irregularly, it was quite difficult even to sense the approach to the top or bottom of a roll.[3] There was, moreover, a noticeable lag between the decision to fire and the moment when a shell left the muzzle. This was partly due to the time it took a gunner to fire once he had decided to do so, but it was also due to the time taken for a powder charge to ignite, and for the projectile to run down the barrel of the gun. During that time, the vertical motion of the gun would be imparted to the shell. Experienced gunners in Nelson's time, for example, used the rolling motion to pitch their balls into an enemy's rigging or against his waterline when the point of aim was on their side.

In the wake of the Napoleonic Wars there were attempts to use a pendulum to detect the appropriate firing point in the ship's roll. The pendulum defined a direction in the earth, and did not require observation of the target or the horizon.[4] In effect this was the beginning of an alternative to Scott's technique, better

Before World War I the rolling ship motion that gunners tried to cancel out moved the gun barrel almost directly up and down, as it pointed along the broadside. Guns pointing closer to fore and aft were affected by trunnion tilt, the effect of which is shown in this diagram from the 1950 edition of the US Navy's gunnery manual. The US Navy called the effect cross-roll, and the fixed stable vertical (Mk 32) introduced in the 1930s made it possible for a ship to fire at a selected point in the cross-roll. By World War II remote power control made it possible for guns to move to cancel out cross-roll altogether. This diagram shows the effect of trunnion tilt and the sort of corrections needed to deal with it. LOS is the line of sight to the target, the line in which a director *not* corrected for cross-level would point a gun. The gun is mounted in the deck plane, and it is aimed with respect to that plane, but fire control is calculated for a horizontal (non-rolling and -pitching) plane. Until after World War I, navies assumed, in effect, that guns would be fired on the broadside, and that correction had to deal only with the rolling motion of the ship that in effect elevated and depressed them. World War I showed that ships would often fire close to the fore and aft line, so that trunnion tilt might be just as important as the up and down movement of the line of sight. Dealing with trunnion tilt required correction in both train and elevation, but train was the more important error.

suited to more massive weapons. The twentieth-century equivalent to the pendulum was a gyro, which tries to maintain its direction in space (not even with reference to the earth). That it should behave so completely against intuition is due to some deep facts of physics.[5] A gyro defined a direction, and thus could fire a gun at a set point in a ship's roll, even if the horizon was invisible. The first patent for a continuously running gyro suited to gunnery was taken out in 1906. The first applications to gunnery were gyro-compasses, which could cancel out a ship's yawing motion while tracking target bearing. The first was patented in 1908. It proved more difficult to use a gyro to define a vertical, which could be used even when the horizon was invisible. Such gyros were precessional, and at least until World War II they had to be corrected periodically by reference to the horizon. Gyros were later the basis for the inertial guidance systems that made long-range ballistic missiles so fearsome: they could sense where they were, without referring to anything external such as the surface of the earth.

The upper limit for continuous aim seemed to be 9.2in calibre (under Scott's command the cruiser HMS *Terrible* doubled the hitting rate for 9.2in guns and more than tripled it for 6in guns). Thus it seemed that fast-firing, medium-calibre guns could outrange heavy guns. They might not be able to penetrate the thickest belt armour, but their high-explosive shells could tear up a ship's side and upperworks, which might be enough to neutralise her. The capital ship of the future might be a fast cruiser. To some extent this was the germ of the battlecruiser idea.

If Scott could make medium guns so effective, what could be done with heavier ones? The potential prize was enormous. Each 12in shell was four or five times as destructive as a 6in shell, and it was more likely to hit at longer ranges due to its flatter trajectory. Because these guns could not be aimed continuously, the Royal Navy sought new methods of firing and gun direction. It also experimented with more and more responsive hydraulic machinery to manoeuvre the guns, the hope being that eventually they could be continuously aimed.[6]

In July 1907 Captain John Jellicoe, the outgoing DNO, reported that new hydraulic engines made it possible to 'hunt the roll', ie, to achieve continuous aim in elevation.[7] In his 1908–9 Estimates Admiral Fisher announced that such gear was being fitted to all British warships with turrets. By 1912, equipped with improved swash-plate hydraulic engines, gunners on board HMS *Orion* could handle a twelve-degree roll (out to out) and some could overcome sixteen- or even eighteen-degree rolls.[8] The existence of such machinery was an important secret, because it made very rapid fire possible, particularly at fairly short ranges. Even when full continuous aim was impossible, an approximation to it made for much more effective long-range control.

Even more important than continuous aim was Scott's emphasis on competition. On the China Station in 1901, instead of the usual competition among the gun crews on board each ship, the ships themselves competed. Captains suddenly became very much aware of gunnery. Officers were encouraged to innovate so that their ships would fire better, and this innovation drove the very rapid progress achieved by the Royal Navy between 1901 and 1914.

Scott's success led to his appointment in April 1903 as the chief gunnery instructor of the Royal Navy, commanding HMS *Excellent*, its gunnery school. From 1905 to 1907 he was the first Royal Navy Inspector of Target Practice, a title emphasising his fleet-wide role. In effect he was chief gunnery-development officer. Scott's appointments undoubtedly reflected the new understanding that the number and character of British warships were not alone sufficient to maintain supremacy at sea. The Royal Navy had to be supreme in gunnery, too. Scott, in turn, inspired William S Sims, a US officer who met Scott on the China Station. Returning to crusade for better gunnery in the US Navy, Sims too became inspector of target practice.

Scott's achievements ignited passionate interest in naval gunnery. Before 1901 *Brassey's Naval Annual*, then the most sophisticated unclassified guide to current naval technology, described in its ordnance section only armour and guns. That year it added a

THE GUNNERY PROBLEM

The initial indication of the gunnery revolution in the Royal Navy was the replacement of earlier fighting tops with fire-control tops. HMS *Britannia* here displays her squared-off fire control tops, with their windows, and the grey colour that replaced Victorian black-and-buff livery. *Britannia* was a *King Edward VII*-class battleship, a transitional stage to HMS *Dreadnought*, with four 9.2in guns in turrets (twin 7.5in turrets were originally planned) plus the usual secondary battery of 6in guns. *Britannia* and *Hibernia* could be distinguished, as completed, by their small, square fire-control tops. The others in the *King Edward* class all had large oval tops on their foremasts, with the usual large fire-control top below on the mainmast. In 1909–10 all but *Britannia* received range-indicator drums for concentration fire.

new chapter on the 'Accuracy and Rapidity of Fire': 'the efficiency of a gun depends to a far greater extent than is generally admitted on the rapidity of its fire or, to put the matter correctly, on the rapidity with which it can hit.' The following year *Brassey's* quoted First Lord of the Admiralty (equivalent to a US Secretary of the Navy) Lord Selbourne: 'gunnery, gunnery, gunnery' is paramount. In 1902 Edward W Harding RMA (Royal Marine Artillery, ie coastal-defence artillery) published a series of articles on 'The Tactical Employment of Naval Artillery' in the 'Traction and Transmission' supplement to *Engineering* magazine. They were collected and published as a book – the first on modern naval fire control – in April 1903.[9] As assistant to the Director of Naval Ordnance (DNO) from December 1903, Harding wrote several important internal publications, including DNO's summary of the gunnery lessons of the Russo-Japanese War of 1904–5.

The need for reach

Scott's gunners proved that they could hit reliably at ranges of about 1500 yards. Anything beyond that

distance required more than reliable gunlaying; range had to be measured accurately and, to some extent, target motion predicted. Greater range became imperative for the Royal Navy just before the turn of the twentieth century because of a new threat, the gyro-controlled torpedo. The gyro kept a torpedo on course while a new engine, using heaters, drove it much further. High-speed torpedo range increased from about 800 to about 1500 yards – to what was then understood as maximum effective gun range. At reduced speed a torpedo could now run for 3500 yards. Torpedoes already armed battleships. Once they outranged guns, they could become primary weapons.[10] Although individual battleships might well find it difficult to aim their torpedoes in the thick of a gunnery action, and although torpedoes were inaccurate at long range, it was widely understood that a line of battleships would be a virtually unmissable target. Moreover, although ships were armoured against gunfire, as yet they had little or no underwater protection.

In 1899 Admiral Sir John Fisher became commander of the British Mediterranean Fleet, at that time the main Royal Navy formation. He was, among other things, a torpedo expert, thus painfully aware of the dramatic improvements in torpedo performance. (He had also commanded the Royal Navy's gunnery school, HMS *Excellent*). Largely to outrange torpedoes, Fisher wanted his fleet to shoot at longer range. He knew that his French opposite number, his most likely wartime opponent, expected to fight at 5000m (about 5500 yards). It is not clear whether he, too, was thinking in terms of the torpedo problem, but at this time the French Navy emphasised the role of torpedo craft. In July 1902 Fisher stated that he would fight outside 4000-yard range to avoid torpedo damage, and, in 1903 manoeuvres, ships caught within 1800 yards were counted as disabled or sunk by torpedo fire. In 1899 Fisher ordered experimental firings at 6000 yards; they were repeated in 1900.[11] Presumably the Mediterranean Fleet experiments explain the Admiralty's decision in February 1901 to order one 6000-yard practice – ie, at four times previous range – per year. Results were poor. The Admiralty concluded that new techniques were needed.

The push for greater range was widely reported. In 1901 *Brassey's* reported on trials conducted the previous year in the Channel and in the Mediterranean at previously unheard-of ranges between 3000 and 7000 yards. 'Without crediting all that has been published of cruisers making thirty per cent hits at a target supposed to represent another cruiser at 5000 yards, it is certain that good practice has been made at ranges never before dreamt of, at any rate for real fighting.' Reliable reports suggested 10 per cent hits at 5000 yards and 5 per cent at 7000 yards.

The torpedo threat continued to drive British thinking on fire control until at least the outbreak of World War I, although it was always much more theoretical than demonstrated. In the one battleship war prior to World War I, the Russo-Japanese War of 1904–5, which strongly influenced naval thinking, neither battle fleet fired mass torpedo salvoes, perhaps because both still used short-range weapons.

Rangefinding

Only once a ship's motion had been cancelled out did it really matter whether the range to the target was known. Attempts at naval rangefinding date back to at least the mid-nineteenth century.[12] All involved some form of triangulation. If the angle subtended by a known distance can be measured, the range can quickly be calculated. The distance can be the length of a ship, the height of a mast, or the known baseline of a rangefinder. By 1855 Sir Howard Douglas, a British writer on naval gunnery, proposed a stadimeter, which measured the vertical angle represented, for example, by a mast of known height on a target. It was compact enough to be carried aloft by a fire-control officer. Unfortunately the user could never be sure of knowing the heights of enemy masts. The alternative was a sextant: an observer atop a mast of known height measured the depression angle between the target's waterline and the horizon. This method was used by some navies as late as 1906. Neither technique was altogether satisfactory, although the stadimeter survived long after World War II as a navigational instrument.

The Barr & Stroud 9ft rangefinder was standard in the Royal Navy before World War I. This one is on the prototype Argo gyro-stabilised mounting adopted by the Royal Navy on its 1909 test aboard HMS *Natal*. The rangefinder needed two operators, one to bring the two images into coincidence and the other to read off the range.
PHOTOGRAPH COURTESY OF PROFESSOR JON TETSURO SUMIDA.

The 1885 Royal Navy gunnery handbook described a two-man rangefinder offered by Major Poore RMA and Captain Pringle, using the length of the ship as a baseline. This idea, which would recur several times, was rejected because it relied so heavily on communication between the two well-separated operators and because it would be ineffective for a target fine on the bow or stern. Even so, with a practical rangefinder in prospect, the Royal Navy became interested in communicating range from it to the guns. By 1885 range telegraphs proposed by Lieutenants Lloyd and Anson were already being tested. Rapid communication mattered because it was understood that in action the range might well change at a rate of 200 yards or more per minute. Nothing came of these experiments.

Expected range (at which battleship guns were set to converge) was 800 yards, and it was widely supposed that seasoned naval officers could estimate ranges well enough (using their 'seaman's eye' as it was customarily described) to hold their fire until they were within decisive range. In 1889, however, another two-man rangefinder was proposed by Lieutenant (later Rear Admiral) Bradley Fiske USN. It was tested successfully by the US and French navies, but never adopted by either.[13] Fiske's experiments impressed the Royal Navy sufficiently to inspire an attempt to develop a British rangefinder.[14] In 1891 the Royal Navy advertised for a rangefinder effective to (ie, accurate to within 1 per cent at) 3000 yards. The following April it tested three devices on board the cruiser *Arethusa*: a 5ft Barr & Stroud horizontal coincidence rangefinder (which they designated FA Mk I), an 8ft vertical Mallock rangefinder, and a two-man Watkin rangefinder (based on the earlier Mekometer in widespread British army service). Fiske's rangefinder was not deemed worthy of sea tests: (his agents protested).

Barr & Stroud won in November 1893.[15] Its descendants were adopted by most of the world's navies. Lenses and mirrors at each end of a tube produced a split image, the top of the target seen through one lens and the bottom through the other. Alternatively, one lens inverted the image. In either case the operator measured the angle between the two mirrors (which gave the range) by matching the top and bottom images, typically by moving one mirror. He relied on a vertical element such as a mast or funnel for his match. Early in World War I the British therefore tried to break up the vertical lines of their masts and funnels with spirals around masts and then with triangular inserts (rangefinding baffles). The British did not know that the Germans were using stereoscopic rangefinders immune to such measures.[16] Even so, most ships had discarded these measures by the time of Jutland. Some apparently never used them at all.

Effective rangefinder range was determined by length and magnification, a typical standard being the range at which the instrument was accurate to within 1 per cent. A 9ft rangefinder could maintain accuracy at greater ranges (but only about 1.4 times as great) as a 4½ft device, but Barr & Stroud found that its longer tubes were not rigid enough to keep the mirrors at each end properly aligned. It found a solution in substituting prisms for mirrors. It also found that, with a fixed baseline, greater magnification improved performance: the features brought together by the rangetaker became more distinct, hence the coincidence more precise.[17] The Royal Navy considered the resulting 9ft FQ2, which it adopted in 1906, accurate to eighty-five yards at 10,000 yards and to 1 per cent (150

The US Navy associated long-base rangefinders with the long battle ranges it sought. When USS *Florida* was completed, she had standard 10ft base rangefinders atop her cagemasts. By 10 December 1916, when this photograph was taken at Hampton Roads, she had much longer base units atop her superfiring turrets. From the *Pennsylvania* class on, long-base rangefinders were mounted in the turrets.

yards) at 15,000 yards.[18] Service performance was considerably worse, however; in 1913 HMS *Thunderer* experienced an average spread between readings from three rangefinders of 700 yards at 9800 yards. In March 1917 tests three ships found errors of, respectively, 1000, 1450, and 1500 yards at 19,000 to 21,000 yards.[19] Factors included refraction and the heating of the rangefinder tube.

The British later paid considerable attention to averaging multiple rangefinder readings in hopes of eliminating random errors. Initially additional rangefinders were fitted in turrets for local control in the event that the main rangefinder was knocked out. They were approved by mid-1912 for the new super-dreadnoughts, beginning with the *King George V* and *Queen Mary* classes. The following year installations in all turrets were approved for all the battleships, although installations began only with the outbreak of war.[20] Although turret rangefinders were associated with local control, by 1913 Fleet Instructions described rangefinder control as the preferred fire-control technique, and pointed out that this method required the maximum number of individual rangefinders (whose outputs were averaged to improve overall accuracy).

In 1907 DNO asked Barr & Stroud for a 15ft rangefinder, apparently to extend precision ranging to 20,000 yards.[21] Tests began in 1909, but results were initially disappointing. In 1912 an improved prototype was being built for installation in B turret of HMS *Ajax* (alternate director position). DNO's priority was to provide armoured control towers for rangefinders atop conning towers, the hope being that this position would be free of smoke.[22] The improved 15ft rangefinder was bought for the *Queen Elizabeth* and

R classes, but not, before the outbreak of war, for any earlier ships. Just why is not clear, as the relevant documentation has not survived.

During World War I the British claimed that the 9ft instrument was good enough out to 16,000 yards or beyond. The US Navy, which hoped to fight at similar ranges, was buying instruments with much longer bases. When its battleships joined the Grand Fleet in 1917, their officers told the British that with such short rangefinders they were 'trying to make bricks without straw.' Although the British saw little merit in most US gunnery practices, they admitted that, given the surprisingly long ranges encountered during the war, the Americans were probably right about rangefinding. After the Battle of Jutland they installed 15ft instruments in their earlier battleships.

In 1893 a German firm, Zeiss, developed a stereoscopic rangefinder in which each lens fed its image into one of the operator's eyes. Unlike a coincidence rangefinder, it had no major moving parts. An operator with perfect binocular vision (both eyes exactly equivalent) saw a single image but with a sense of depth. He found the range by moving a marker (the Germans called it a 'wandermark') until it coincided with the target. Thus it was possible to range on an object of irregular shape, such as a shell splash. This

Coincidence rangefinding required a vertical line on the target, on which the operator could make a 'cut'. During World War I the Royal Navy tried to break up the vertical lines of its masts and funnels with canvas 'baffles,' shown here on board HMS *Barham* at Scapa Flow in 1917. These measures turned out to be pointless because, unknown to the British, the Germans used only stereo rangefinders.

At Guantanamo Bay in 1920, USS *Pennsylvania* shows typical post-World War I modifications, such as the addition of enclosed mast platforms for torpedo defence, and an enlarged bridge topped by a large rangefinder. No such device could occupy the roof of B turret, because it had been adapted for a flying-off platform – an early approach to air observation. Both the enlarged bridge and the big rangefinder testify to World War I experience with the British Grand Fleet, which had introduced tactical plotting as a way of maintaining what would now be called situational awareness. The extra rangefinders were usually called tactical; they were a way of measuring ranges to nearby ships so as to maintain an accurate plot of the evolving tactical situation. Such practices were necessary if complex tactical evolutions were to be carried out. Note the turret rangefinders introduced in this class.

After World War I, exposed US rangefinders were given splinter shields. This one is aboard USS *Wyoming*, shown firing sub-calibre practice off Panama, 10 March 1926. Note the two openings in the center of the rangefinder, for the (target) finder and the pointer who kept the rangefinder on target. Once the device was on target, the finder read off the ranges while the pointer kept the two halves of the coincidence image matched, to keep the rangefinder at the appropriate range. In later rangefinders the finder had a telescope whose aperture was at one end of the rangefinder. During World War I the US Navy found that forward turret-top rangefinders were all but useless in a head sea; in February 1918 Commander Battleship Division Nine (the ships with the Grand Fleet, designated as the 6th Battle Squadron in that fleet) pressed for relocating the rangefinders to a protected position above and clear of the conning tower. The turret markings indicate the bearing to which it is trained. They were introduced during World War I by the British Grand Fleet as an aid to concentration firing by several ships against the same target (range dials were introduced at the same time). By 1926 such markings were rare in the US Navy.

technique was considered more sensitive, hence more effective in poor light, but it also demanded more of the operator. A naval version was advertised in 1906, and the Germans officially adopted it in 1912.[23]

The British later argued that few potential operators had sufficiently acute stereo vision, and that their vision would likely deteriorate with fatigue or battle stress. During the 1930s several navies, including the US, French, and Italian, followed the Germans in adopting stereo rangefinders. British contact with the World War II US Navy changed their attitude; by 1943 they admitted that stereo was not only acceptable, but better in poor visibility.[24] The US Navy had had no problem at all in finding operators, nor did it experience the supposed battle-fatigue problem. Stereo units were already clearly superior for anti-aircraft fire, as an airplane might present no feature on which a 'cut' could be taken. The new British policy was to fit ships with 50 per cent stereo rangefinders, except for close-range units and ships with only a single rangefinder. Contracts were let to two British firms and to some US firms; as of 1943 the first units were scheduled to go to sea in 1945.

The rangefinder measured the *geometric range* between shooter and target, which was sometimes called the true range. Given rangefinder errors, this measurement could not be entirely accurate, but it is convenient to identify the rangefinder figure with the actual distance between ship and target. This range was not the same as the *gun range*, the range to which sights should be set. Gun range took into account the movement of the target while the shell was in the air and even that of the shooter while the shell was in the gun (where it shared the ship's motion). It thus involved knowledge of how the range was changing: the *range rate*. The longer the range (ie, the more time the shell spent in the air), the more significant the range rate. At very long range, factors such as the

After reconstruction in the 1920s, early US dreadnoughts retained their external rangefinders. Note the massive external rangefinder atop B turret. USS Arkansas is shown at Norfolk Navy Yard, 27 June 1942, just having been modernised again (she had retained her cage foremast after her 1920s modernisation). The new tripod foremast was needed because the old cage mast could not support the radars (SC air-search and Mk 3 fire-control radars) ordered installed in mid-1941. Despite her radically altered appearance, with triple-decker tops as in later reconstructions, she retained her original surface fire-control system (Mk 2 Mod 4 directors in the tops for her main battery, a Mk 4 Mod 1 in her conning tower, and two Mk 7 Mod 10 for her 5in/51 secondaries). Her Mk 9 stable element was replaced by a unit removed from USS Tennessee. In 1944 it in turn was replaced by a Stable Element Mk 6 that could stabilise the Mk 3 fire-control radar. Despite wartime requests, the ship's non-synchro system was never replaced, and her commanding officer complained of numerous breakdowns off Okinawa in the spring of 1945. Arkansas and her sister ship Wyoming, converted into a gunnery training ship under the London Naval Treaty of 1930, were the only US battleships with 12in/50 guns.

rotation of the earth had to be taken into account. It began to matter that a ship was able to measure her own speed. That was difficult: only in about 1912 did the Royal Navy obtain an electric log (measuring speed). Other navies were probably in about the same position: the Germans license-produced the British log.[25]

The British were fairly sure, moreover, that their understanding of gunnery was far in advance of any other navy: in 1906 DNO's assistant Captain Harding remarked that foreign navies did not yet understand the difference between geometric and gun range.[26] Only recently had British officers realised how important it was to know the geometric range precisely, rather than depend on spotting beginning with an approximate range. Presumably this referred partly to Captain W C Pakenham's comments during the Russo-Japanese War (Pakenham was Royal Naval attaché to Japan at the time): 'Outside the Service the impossibility of continuous use of the rangefinder and therefore the importance of a knowledge of the rate of change [range rate] is not recognised, consequently the means of its determination are *unsought for*.' No one had tried to make a rangefinder record its output automatically, and no one (apart from Pollen, see chapter 2) had realised the importance of using a gyro to eliminate yaw from rangefinder bearing readings. The Germans were probably the most advanced foreign navy at this time. Little was known of their thinking, but the evidence of what they were using (sextants with a few unmodified Barr & Stroud rangefinders) and of articles in their main annual publication, *Nauticus*, suggested that they were not working along British lines.

The range rate

Successful gunnery required that the position of the target be projected ahead, ultimately to the moment at

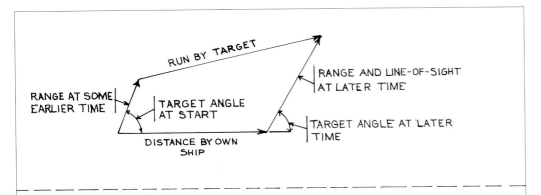

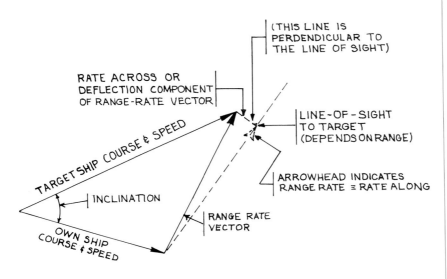

As a gunnery lieutenant, John Saumarez Dumaresq made a fundamental discovery: that the range rates (across and along) did not depend on the range, only on target course and speed. The top diagram shows the gunnery problem (it did not matter whether the target is moving towards the shooter or away from it). The lower diagram shows the way in which the rates at which own ship and target move (their course and speed) give the rate (a vector: speed and direction) at which the range is changing – the quantity the fire-control system needs. The Dumaresq gives the two components of the vector, the rate along (in the direction from shooter to target), which was also called the range rate, and the rate across (deflection – but in knots, not in terms of degrees of bearing). The Dumaresq modelled the situation, using a bar pointing at the target to select the range-along component of the range rate. Own-ship and enemy-ship bars modelled the vectors of own-ship and enemy-ship course and speed. The operator had to estimate inclination (the angle between own and enemy course) and enemy speed. Dumaresqs and their equivalents were the basis for the later mechanical fire-control computers, because they enabled the computers to translate enemy-course and speed estimates into rates that could be integrated to give range and bearing.
(A D BAKER III)

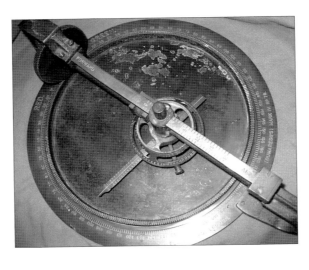

As a Rear Admiral, Dumaresq was Commodore of the Australian fleet (he had been born in Australia), having commanded HMAS *Sydney* during World War I. The brass Dumaresq prototype is in the Royal Australian Naval Historical Collection. Donated by Admiral Dumaresq's daughter, this Mk IV survives in the Royal Australian Navy Heritage Centre. The large bar represents own ship, the small one the target. The cross-piece that connects the own-ship-speed pin to enemy speed to indicate the rates is missing, as are the scales on the face of the circular plate.
(RAN SEAPOWER CENTRE)

which a shell might be expected to hit. To do that, the shooter had to calculate the rates at which the range and bearing of the target changed; they were usually called the range and bearing rates. Calculation was difficult because neither was constant, and because each depended on the other. Alternatively, one might think in terms of the vector (magnitude and direction) pointing from shooter to target. The change in this vector was another vector which might be called the rate vector. It could be expressed as two components, one along the line of fire and one across it. The rate along was usually called the *range rate*. The rate across was usually called *deflection*. Its magnitude was the bearing rate multiplied by the range.

Fire-control systems contended with numerous time gaps and dead times, for example between rangefinder cuts (observations), or between observation and sight-setting and firing. Knowing the range rate made it possible to bridge time gaps between observations, or between observation and sight-setting, or to correct aim for the next salvo based on splashes from the last, while shooter and target moved. Prediction came to be called *range-keeping* or *position-keeping*.

In 1902 a gunnery officer, Lieutenant (later Rear Admiral) John Saumarez Dumaresq RN, made a remarkable discovery. The rate *vector* (magnitude and direction) between two ships moving at constant speeds along steady courses did not change over time. It did not depend on range. What did vary were the *components* of the vector along (ie, the rate along or range rate) and across the line of fire (the rate across), because the direction from shooter to target changed as both steamed along. If the rates were changing slowly, a graph of either against time would be a nearly straight line. This approximation was the key to the Dreyer Table described in chapter 2. The rate at which the range rate varied depended on the rate across, and vice versa.[27]

Dumaresq designed a simple physical analogue of the engagement.[28] Separate bars represented own-ship and target. The target bar was pointed along the target course, pins indicating own and target speeds. A dial with a grid engraved on it was turned so that a third bar pointed along the line of sight to the target. The target bar then pointed to the appropriate range rate and deflection (in effect, two components of the relative target speed) on the engraved grid. Of course, enemy course and speed could not be measured directly; they had to be inferred. At short ranges, however, it was relatively easy to guess enemy course by how foreshortened the target looked, and speed might be estimated from the appearance of the enemy's bow wave. The analogue (or Dumaresq as it was called, after its inventor) became a key British gunnery instrument. Ultimately course-solvers were key components of the analogue computers of the mature surface fire-control systems. Dumaresq patented his device in 1905. Analogous course-solvers appeared in other navies.

The Dumaresq gave *speeds*, in yards per minute or in knots, both along and across the line of sight. Although guns had deflection sights (to lead or trail the target) marked in knots, in fact the enemy's speed across the line of sight could not be measured directly. What an observer saw was a change in bearing, ie, of the *angle* to the target. The speed across is the bearing rate (angle) multiplied by the range. The longer the range, the slower the *angular rate*: distant objects seem to move slowly. Later it was useful to translate between Dumaresq output and observable bearing rate. This generally took a human operator using a slide rule or extrapolating from a curve, either of which took time and potentially introduced errors. Some Dumaresqs had additional cross-lines giving bearing rates (angles) for different ranges, so that, given a deflection rate, the bearing rate in degrees could be read off (or estimated). It took a computer to translate smoothly, using a range carried in its analogue mechanism.

In 1913 a version of the Dumaresq was connected to a gyro so that own-ship manoeuvres did not disturb the target bar. The line-of-sight bar *at the moment of manoeuvre* was also correctly moved. However, the line of sight was moving all the time. The gyro connection did nothing to keep it moving with the target. That took continued observation of the target. In the Dreyer Fire-Control Table (see chapter 2), this modification went part of the way – but hardly all the way – towards making helm-free firing possible.

The alternative to the Dumaresq was to measure the range rate directly. There was no alternative means of measuring deflection or bearing rate. The 1904 committee, which in effect invented British fire-control practices suggested that successive ranges be read off against the times they were taken, using a stop-watch. The difference between two such ranges was the range rate. This very straightforward approach was tested against a Dumaresq. To the obvious surprise of those conducting the test, the Dumaresq worked far better. Random errors were the problem. Range rates were small, and rangefinders often failed to register them. Given estimated enemy course and speed, and actual enemy bearing, the Dumaresq produced range and bearing rates. They could be used to project forward estimated target range and bearing, based on initial data. Comparing projections with observation tested the estimated rates and, by extension, the initial

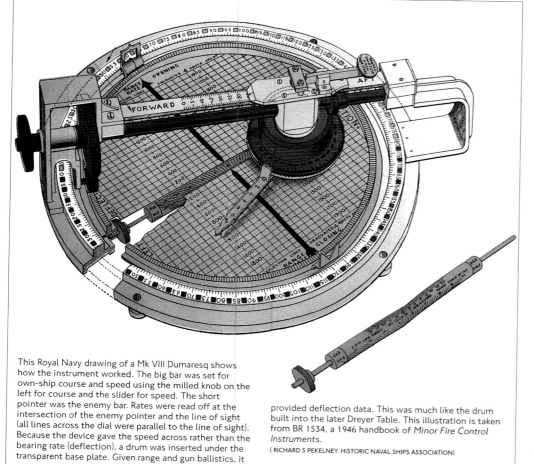

This Royal Navy drawing of a Mk VIII Dumaresq shows how the instrument worked. The big bar was set for own-ship course and speed using the milled knob on the left for course and the slider for speed. The short pointer was the enemy bar. Rates were read off at the intersection of the enemy pointer and the line of sight (all lines across the dial were parallel to the line of sight). Because the device gave the speed across rather than the bearing rate (deflection), a drum was inserted under the transparent base plate. Given range and gun ballistics, it provided deflection data. This was much like the drum built into the later Dreyer Table. This illustration is taken from BR 1534, a 1946 handbook of *Minor Fire Control Instruments*.
(RICHARD S PEKELNEY, HISTORIC NAVAL SHIPS ASSOCIATION)

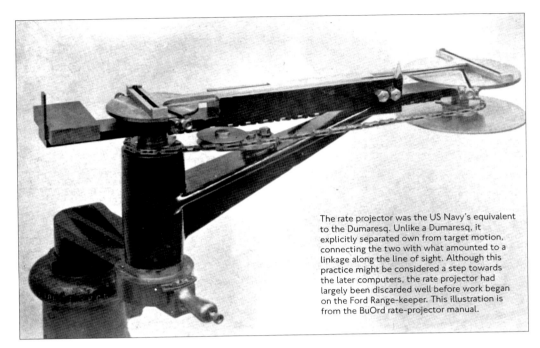

The rate projector was the US Navy's equivalent to the Dumaresq. Unlike a Dumaresq, it explicitly separated own from target motion, connecting the two with what amounted to a linkage along the line of sight. Although this practice might be considered a step towards the later computers, the rate projector had largely been discarded well before work began on the Ford Range-keeper. This illustration is from the BuOrd rate-projector manual.

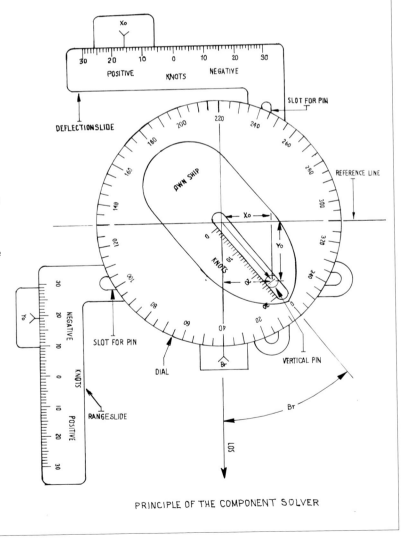

Dumaresq's idea survived into World War II fire-control computers, but typically own and target motion were separated. The US Navy's component solver, illustrated here, was essentially a Dumaresq, displaying the range rate (vertical slide) and the rate across (deflection slide, horizontal). In effect the motion of the other ship was set to zero (a separate dial gave data for the other ship). The positions of these slides in turn represented quantities that could be added and multiplied to produce range rates for integration. The relationship between rate along and rate across is set by the pin in the dial showing speed and bearing (Br). A range-keeper had two such dials, one for target and one for own ship, set so that the line of sight (LOS) connected them. The target dial in turn rotated as the fire-control solution was generated. (A D BAKER III)

assumptions as to enemy course and speed which had gone into the Dumaresq. This technique worked.

Transmitting data

The range to the target had to be transmitted to the guns. Using voice, via either a pipe or a telephone, invited errors. It was never certain that what went onto the sights corresponded to rushed digits heard through the noise of battle. However, any other mode of transmission required considerable ingenuity in an analogue age unlike our own. Making data transmission work became a major issue in fire-control development because otherwise the rest of the effort was wasted. Most techniques were electrical.

In about 1894 Barr & Stroud offered the first practical approach to this problem. A pointer on a dial ran by clockwork as long as the current was on. That ensured that the pointer moved at a constant speed: its position on the dial, indicating a range, was proportional to the duration of the power pulse. Moving the transmitter to the position to be indicated turned the current on, then off. The British 1904 firing trials showed this method to be unreliable. Nor could the dial be interpreted quickly enough. The trials board wanted counters (cyclometers) – as in a car's speedometer – from which numbers were read directly. But the fleet urgently needed data transmitters. While it shopped for something better, the Royal Navy urgently bought a Barr & Stroud Mk I with two dials, transmitting range in the new standard twenty-five-yard steps.

Vickers solved the problem with a step-by-step electric motor that could drive a cyclometer. Step-by-step meant that the rotor was turned a set amount by each power pulse. Energising one or both of the two sets of rotor windings turned the rotor a quarter-turn (each pole position had separate wiring, for a total of five wires). The transmitter energised the receiver windings in sequence, turning the rotor with it. As a safety feature, the stator windings were energised through the check-fire switch. The receiver turned a cyclometer wheel. When the first row of numbers (00, 25, 50, 75) turned over, the second began to turn, and so on.

Although fire control was clearly important, exactly what it required was long a matter of controversy. In 1908, when the *Colossus* class was being designed, fleet commander Vice Admiral Sir Francis Bridgman argued that masts themselves were a danger, and that it would be better to build mastless ships and accept reduced accuracy. Several foreign navies followed that advice, avoiding aloft fire-control positions. DNO Captain Reginald Bacon replied that aloft control would be particularly important during the approach phase of an action (he thought that once the action became general, rangefinding and formal control would no longer be very important). Bridgman's ideas led the Royal Navy and others to provide armoured fire-control stations close to the conning tower, such as the spotting position under the bridge of HMS *Indefatigable* (her sisters did not repeat it). The heavy mainmast was considered useless for fire control because it suffered so badly from smoke interference. Smoke interference with the forward control position aloft was no longer considered as important as the armoured position below. Nor did battleships need a mainmast to support radio aerials (although battlecruisers, which needed greater radio range, retained tall topmasts). This logic (and probably also the advantage of preventing an enemy from gauging course over the horizon) revived the foremast arrangement used by HMS *Dreadnought* and then abandoned. The single mast was placed abaft the forefunnel because that way the vertical member of the mast provided a support for the heavy boat crane; in classes with two tripods, the vertical member of the mainmast supported the crane. This arrangement was first used in the *Colossus* class. Designed before those ships went to sea, hence before severe smoke interference had been demonstrated, the first 13.5in gun ships (*Orion* and *Lion* classes) had similar foremasts, although the battlecruisers also had mainmasts. *Lion*, shown as built, experienced particular problems because of her great boiler power even though, as in the *Orions*, the forefunnel was fed by a reduced number of boilers. After preliminary trials in January 1912 Captain A A M Duffy wrote that control from the top 'would be of little or no use in war'. The spotting tower, which had been seen as a viable alternative, was obscured by the bridge whenever the ship rolled more than five degrees, which was usually the case. The Admiralty was already aware of the problem, having called a fire-control conference in November 1910 after deciding the previous April to buy the stabilised Argo rangefinder. The *Lions* were soon rebuilt at considerable cost. The conference decided to provide future ships (*King George V* and *Queen Mary* onwards) with an armoured control position aft instead of the spotting tower atop the conning tower. After the necessary space had been provided, but before ships were completed, this position was abandoned in favour of using B turret as secondary (alternative) control for the main battery. The after structure was used for torpedo control. It also supported a second main 9ft rangefinder. These functions continued through at least the *Queen Elizabeth* class. This caption is based partly on John Brooks, 'The Mast and Funnel Question,' in *Warship 1995*.

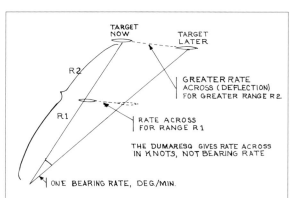

What an observer sees is the bearing rate of the target, the rate at which the angle to the target seems to change (which has to be corrected for the fact that both own ship and target ship are moving). However, what a Dumaresq or its equivalent gives is the speed across, in knots. The speed across depends on the range, so it is impossible to compare the output of a Dumaresq directly with observation. Any fire-control device has to take the range into account.
(A.D. Baker III)

Vickers' transmitter-receiver pairs equipped forty-two ships, beginning with HMS *Dreadnought*. The technique was slow, but it became significant because it was easy to extend to other kinds of displays, such as a follow-the-pointer dial at a gun.[29] The Mk II version used for follow-the-pointer in effect turned the original version inside out, the current being applied only to the six fixed magnets of the stator, energised in pairs (energizing two pairs in sequence turned the rotor). The receiver motor was geared to the range pointer, moving it fifteen minutes of arc for each turn of the transmitter.[30] The dial was placed on the sights. The operator copied data by moving a pointer on an outer dial so that it matched the transmitted number on the inner dial. Follow-the-pointer was far more reliable than having a gunner use the reading on a counter to set his sights. The Royal Navy selected this method after several attempts at automatic sight-setting failed.

If two magnets could fix a rotor within a quarter-turn, more could set it at any desired position. The British Siemens Company, a branch of the German company responsible for German fire control, was the first to offer such a stepping motor. Ten electromagnets were arranged around its stator. Each of ten spring-loaded pins in the receiver could energise three adjacent magnets. Each such combination turned the rotor to a unique position. Fast operation demanded complexity: twenty-eight wires were needed (two for the rotor).[31] The Mk II version (five magnets) needed only seventeen wires. Siemens receiver-transmitter combinations were installed on board nineteen British pre-dreadnoughts. Barr & Stroud beat out Siemens with a Mk-II system using three magnets and a rotor with three soft iron elements. The rotor could line up with a magnet or it could be forced to lie between two of them. Energising the magnets in the right sequence turned the rotor through 180 degrees. It could also be reversed. No wire went to the rotor, hence there was no brush to wear out. The motor had to skip three steps before it became permanently out of step with the transmitter (that this had to be said suggests the sort of problems other transmitters were showing). Introduced in 1907, by 1912 this Mk II was the fleet standard except for follow-the-pointer operation.

For greater precision and instant transmission, Evershed & Vignolles offered a bearing indicator using the electrical balance between transmitter and receiver. Each had a pointer moving along a graduated resistance. Current flowed between them until the resistances were the same. In this way a range of 200 degrees could be covered (if it were 360 degrees, zero and full scale would equate to the same resistance, which was unacceptable). Systems were offered both with one-way and two-way transmission. They generally had two alternative 180-degree scales.[32] Such indicators were first used for helm, and in 1907 the firm offered them for range, deflection, and gun orders. They were considered too massive, but the concept was attractive for target indication. The prototype was installed on board HMS *Superb* in 1910, and the first production instrument installed on board HMS *Bellerophon* in December 1912.[33] 'Eversheds' became particularly important during World War I as a means of ensuring fire-control coordination.

Putting a system together

In the spring of 1904 special fire-control trials were conducted on board the battleship HMS *Victorious* of the Channel Fleet and HMS *Venerable* of the Mediterranean Fleet. The key conclusion was that a ship's guns had to be handled in a unified way, under

THE GUNNERY PROBLEM

Photographed just before the outbreak of war in 1939, the German 'pocket battleship' *Graf Spee* shows her forward 10.5m (34.5ft) rangefinder atop her tower mast. This was not the ship's director; the Germans favoured separate periscopic directors with gyro-stabilised optics. The tower mast did provide the gunnery officer's observation platform, and during the Battle of the River Plate Captain von Langsdorff occupied it (he was criticised for getting in the way of his gunnery officer). He had been trained as a torpedo officer rather than a gunner, and there was speculation that he was too concerned with evading the cruisers' torpedoes and therefore did not steer the sort of steady course that would have made his gunnery more effective. The multiple tubes (sunshades) emerging from the rangefinder indicate that it is actually four instruments in one: a full-size rangefinder, a full-size stereo spotting glass, and shorter-base instruments for the pointer (layer) and trainer. The spotter and range-taker sat on one side, the layer and trainer on the other. This arrangement was typical of German large-ship rangefinders, although in most cases the lenses did not have the sunshades visible here. In *Bismarck* and other capital ships the pairs of lenses were canted rather than being placed one above the other, but there were the same four sets of two lenses each. These details are from a wartime Admiralty Research Laboratory report (ADM 204/635 of 12 July 1945) describing a 10.5m (34.5ft) naval rangefinder captured near Ghent (used for coastal defence; a typical capital-ship type). This unit had magnifications of 18X, 25X, 36X, and 50X. The spotting element had a magnification of 15X; one of its eyepieces had vertical and horizontal lines etched onto it. In this unit the spotting scope had an 11m (36ft) baseline and the layer's and trainer's scopes had a 6m (19.6ft) baseline (magnification 15X). At minimum magnification the field of view of the main rangefinder was two degrees, compared to four for the layer's and trainer's scopes. The range scale in this instrument ran from 3000 to 10,0000 metres. The rangefinder eyepieces had neutral and red filters (the latter for ranging into the sun). Heavy cruisers such as *Prinz Eugen* used 7m (23ft) rangefinders with similar features. Just forward of the masthead rangefinder, and barely visible, is a periscope director; the big rangefinder mount was not a DCT. The object at the base of the mast is an anti-aircraft director. Another long-base rangefinder is visible to the right.

Battle of the Yellow Sea a decisive victory. 'The war marks the highest achievement of a system of gunnery which has now practically passed away' (DNO assumed that Russian practice was broadly similar to Japanese). The Japanese had done better than might have been expected thanks to their remarkable fire discipline and morale. However, even they had their limits. By the end of the drawn-out Battle of the Yellow Sea, their gunlayers complained that the targets looked blurry. DNO doubted that other navies would do as well.

DNO was unaware that the Russians had a centralised fire-control system similar to what he advocated. Japanese medium-calibre fire tore up cables and killed exposed rangefinder parties. The gunlayers had not been trained to fire independently. In effect the Russians provided a cautionary example against what the Royal Navy was doing.[38]

The war demonstrated how difficult it might be to designate targets in the confusion of battle. The Japanese became confused when the Russians 'bunched' and when ships became enveloped in their own gun smoke. With no way of quickly shifting targets, they ceased fire while a new target was designated by word of mouth. That delay might well have been fatal had the Russians been more efficient. As it was, concentrating on pre-designated targets sacrificed key tactical virtues: surprise, containment, mutual support. For DNO, the failure to develop flexible target designation exemplified the way in which technique (fire control or gunnery technology) dominated (and ruined) tactics.

CHAPTER 2
Range-keeping

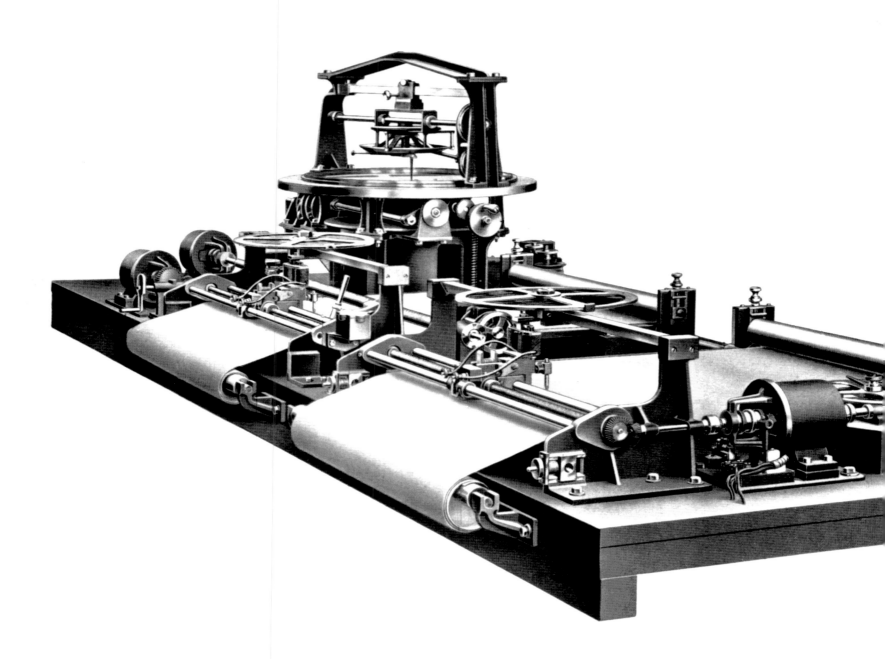

IN LATER US NAVY TERMS, the two alternative approaches to position-keeping were analytic and synthetic. The *analytic* approach was to deduce the range rate directly from observation. It was the straightforward method, but was vulnerable to errors of observation. The alternative *synthetic* approach begins with an estimated solution, then refines it by comparing its predictions (eg, ranges) with measured reality. It turns out to be much better than the analytic method because erroneous data are easier to discard. The Dumaresq was a forerunner of the synthetic approach, and in 1904 it was clearly better than the analytic one. The Dreyer Table was the high point of the analytic approach.

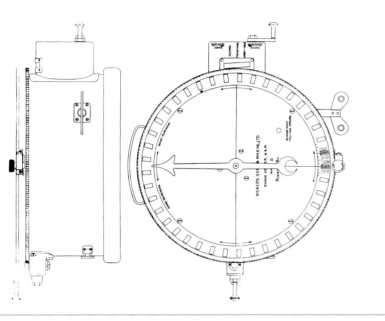

The Vickers Clock was basic to early dreadnought fire-control systems. This drawing of the face of the clock is from the US Navy 1908 manual for what it called the range-keeper Mk II; later the term range-keeper meant an analogue computer, and the Mk II designation was re-used for the 'Baby Ford'. The clock could handle a speed range of 1.5–38.6 or 2.3–60 seconds for a fifty-yard change of range.

The Vickers Clock

The Royal Navy used the Vickers Clock to calculate the changing range based on a set range rate. It was set at the current range and estimated range rate, its pointer moving to show the corresponding estimated current range at a later time. The clock contained a wheel spinning at constant speed. At that speed in revolutions per minute, the rim of the wheel is moving faster (in linear terms, eg, inches per second) than a part closer to its axle. A ball held against the wheel spins at a rate set by the *linear speed* of the place it touches – faster near the rim, slower near the centre. In the clock, this ball, on a shaft at right angles to that of the constant-speed wheel, drove the pointer showing current range. The position of the ball along the face of the spinning wheel was set according to the range rate. The wheel-ball combination was called a variable-speed drive.

Initially all concerned believed (incorrectly) that for two ships on steady courses at steady speeds, the range rate was constant. The clock was not therefore designed to be reset while running. It was no more than a way of avoiding mental arithmetic (the multiplication of the range rate by the dead time) that was subject to error. Gradually it became clear that rates could change (and could be wrong). Clocks did need to be reset, typically once a minute (it was difficult to get revised rates more quickly than that either from Dumaresq or plot). This practice of 'joggling' or 'tuning' wore down the ball, making for slippage that impaired accuracy. If the movement was too rapid, it could be jammed into the disc. The Mk II version had

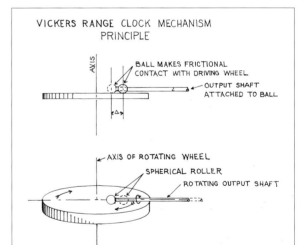

The Vickers range clock was the first gunnery integrator. It converted a position across into rotation at a variable rate, which could drive a clock or a counter. The clock or counter showed how far the output shaft had rotated, in effect adding up the effects of the variable rate. Such clock mechanisms were also called variable-speed drives. This simplified diagram shows how it worked. The wheel rotated at constant speed, so that the rate at which the output shaft rotated depended on where its roller touched the wheel. In a Vickers Clock, the output shaft was geared so that it drove a series of shafts at right angles, so that the face of the clock could be parallel to the driving wheel. In the Dreyer Tables, the output shaft of the same mechanism drove a long screw, which in turn moved a pen across a roll of plotting paper. The main problem in such integrators was that the ball could slip instead of turning. The Barr & Stroud equivalent used a hardened-steel wheel held down against the spinning constant-speed wheel. It was used by the Royal Italian and Imperial Japanese navies.
(A D BAKER III)

a clutch, so that the ball could be disengaged and moved without stopping the constant-speed disc. Disengagement imposed a time lag, however, which could prove excessive if the range rate was changing rapidly.

The clock was what would later be called an *integrator*, adding up increments of range to estimate current range. This use of the word was broadly equivalent to the way integration is defined in calculus. Integrators became key to synthetic fire-control systems.[1]

Proposed in 1903, the clock was widely described and marketed, becoming the common currency of naval fire control by about 1906.[2] The Royal Navy chose to put the clock not in the fire-control top on a mast, but in the transmitting station. It came to use the word 'clock' to mean a fire-control computer. For example, the Pollen computer introduced in 1912 was called an Argo Clock (it was made by Arthur Pollen's Argo company). Based on the later Dreyer Table, the Royal Navy referred to any combination of computer and automated plot as a table, as in the Admiralty Fire-Control Table. In these terms the US Ford rangekeepers were 'tables'.

The output of the clock was present range. By 1913 clocks usually had a second hand, fixed in relation to the main hand, to show a fixed correction for corrected gun range. The component for target movement was usually applied as a spotting correction. For example, the 1913 Fleet Orders suggested that half the range rate, ie, target movement for a shell in flight for half a minute, be used. This technique was unsatisfactory because it required frequent resetting of the second hand, and because it required calculation of the contributions of own- and target-ship movement (by 1907 slide rules for both functions had been devised).

By 1907 Captain J T Dreyer RA (Royal Artillery), the brother of Lieutenant (later Vice Admiral) Frederic Dreyer RN, who invented the principal Royal Navy World War I fire-control calculator, had developed a range corrector giving the total correction for range rate, ballistics, and wind speed.[3] It was placed in the transmitting station; other corrections could be applied directly on the sight (which Dreyer had also developed). The object was to eliminate any need for manual calculation. The input for target movement was an approximation based on a particular range rate multiplied by the time of flight associated with the set range (this calculator was not connected directly to any device measuring the range rate). Presumably the Admiralty had the corrector in mind when it ordered the range-correction element of the Pollen (Argo) Clock eliminated as an unnecessary complication. That the relatively static correction offered by the Dreyer corrector was considered good enough suggests that range rates were expected to change slowly.[4] By 1918 the corrector was deemed too slow; rates were much higher than anticipated, and allowable errors (danger spaces) much smaller, due to longer ranges.

Deflection was also affected by target motion and by wind. Before the invention of the Dreyer Table and the Argo Clock, the effect of target motion was taken from a plot, a bearing rate being applied to the expected time of flight of the shell. Wind speed across the line of fire was measured directly.

Plotting

The idea that plotting was essential to successful naval fire control seems to have originated with Arthur Hungerford Pollen, a civilian who became very influential in pre-1914 Royal Navy fire-control development.[5] He was that typically British creature, a brilliant and very persistent amateur. He headed the Linotype Company, a precision manufacturer but, perhaps significantly, not one making instruments, hence not drawn into gunnery development. However, he was neither an industrialist nor an engineer; he had gained his position at the company by marriage. The Linotype connection gave Pollen access to excellent engineers (Harold Isherwood and his assistant D H Landstad) and to a leading British physicist, Lord Kelvin, who was on the company's board. Pollen had important connections in the British establishment, and he did not increase his popularity with British naval officers by using them to promote his system. It is easy to overstate their significance. His connections did not necessarily mean that he would be taken seriously, as

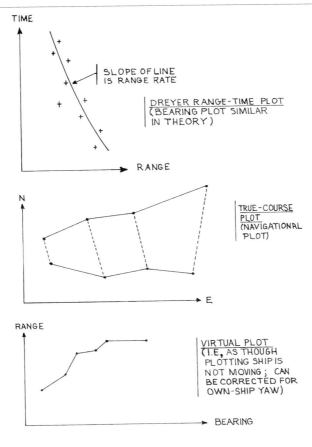

Navies tried three different kinds of plotting to project ahead target range and bearing. One method was to separate out range and bearing in the hope that although they were actually interconnected, that connection would be relatively weak. This was Dreyer's concept. A range-versus-time plot (which the US Navy called, simply, a plot) is shown. The alternative (tracking, to the US Navy) was to try to reproduce the actual motions of shooter and target. In this sketch, the dashed lines represent observations of range and bearing by the shooter (the lower line); the straight lines between observations are estimates of motion. Pollen tried to mechanise this type of plotting. It became easier once good gyro-compasses became available. The third type was a virtual-course plot, as though the plotting ship was not moving. It turned out to be the worst of all, because own and target motion could not easily be disentangled.
(A D BAKER III)

Range-keeping was the basis of fire control: predicting where the target would be when shells had to arrive. Prediction required that the target follow a straight course at a constant speed. Only the very last mechanical computer, the Admiralty Fire Control Table Mk X in HMS *Vanguard*, made any attempt to deal with a manoeuvering target. Before that, the mechanical computers offered a way of dealing with own-ship manoeuvers (by separating own-ship from target motion in their calculations), so that a ship could manoeuvre while hitting. When two ships with such systems engaged, as in the Komandorski Islands in 1943, the results could be entirely indecisive, because both ships could manoeuvre freely. Systems began with what could be measured directly: target bearing angle and target range. Since they could be measured at intervals, as shown, the rates at which they changed could also be measured. That was not enough, because the rates varied over time (and were interconnected). After World War I several navies bought inclinometers, which tried (with limited success) to measure target inclination, ie, course, range, bearing rate, and inclination together gave target speed. Present range and bearing are where the target is right now. The gun has to be pointed ahead of the target (deflection) and aimed at a different (advance) range to hit. The official Italian fire control handbook published in 1933 described a series of six alternative pairs of data that could be used for prediction. Which were best depended on the circumstances.
(A D BAKER III)

he clearly was. Ultimately what mattered was analytic talent and determination.

Pollen became interested in fire control after observing a Mediterranean shoot off Malta by HMS *Dido* in February 1900 at the invitation of his cousin Lieutenant William Goodenough (later a successful World War I cruiser commander). He asked why the ship was firing at only 1400 yards, when similar guns brought ashore were firing at 8000 yards against the Boers in South Africa. Goodenough blamed inadequate means of rangefinding (the few Barr & Stroud instruments in British service were disliked and distrusted).

To visualise the gunnery problem, Pollen plotted the paths of two fast (twenty-five-knot) ships approaching each other on opposing courses. He was impressed by how rapidly the range between them changed: later he claimed to have been the first to understand the significance of range rate (his scenario entailed the highest possible range rate). Dumaresq had not yet invented his rate-finder. Pollen saw plotting as a means of visualisation and calculation; future range might be calculated by extending ahead plots of own and target course. This was true-course plotting. A variant, in which target ranges and bearings were plotted as seen from a ship (ie, plotting as though the firing ship was stationary), was called virtual-course plotting. Neither was yet practicable, because magnetic compasses were grossly inadequate (they could not react quickly enough to compensate for yaw). Because he was not an engineer, and because he had not been to sea, Pollen was unaware of such limitations.

Plotting could have a very different function. Lieutenant (later Vice Admiral) Frederick C Dreyer (in 1908 assistant to DNO Captain Reginald Bacon) observed by 1908 (when he probably became Pollen's rival) that under many circumstances the range rate would be nearly constant and thus the plot of range versus time nearly a straight line whose slope was the range rate. This rate could be applied to a clock.[6] It could be corrected from time to time as the clock was seen to be running faster or slower than the actual range. Dreyer considered this approach far simpler than Pollen's automated one.[7]

Dreyer's real contribution, dating from January 1911 at the latest, was to realise that a plot could solve the problem of random errors. Dreyer's experience had taught him that errors could never be eliminated, and that people were the only hope of overcoming them. By drawing a line through the scatter of readings a human plotter could see and reject bad data (this is now called data smoothing). It was much easier to smooth a pure-range plot than a virtual or true-course plot, in which even correct data might not produce anything like straight lines. Thus plotting made the analytic approach rejected in 1904 much more workable. The straight plot could never be more than an approximation. The rate at which the range rate itself changed depended on the bearing rate, and vice versa, so that neither range nor bearing rate could be treated in isolation. The interaction was particularly strong for high range and bearing rates, as when two ships raced towards each other. Pollen therefore argued that no pure-range plot could suffice for long. Dreyer understood that the range rate could not be constant, but he justified his approach as a useful approximation (he may have imagined that he could extract useful data from a curved plot).[8]

Dreyer's argument was unintelligible to Pollen, who tended to expect that better equipment, particularly using gyros to overcome errors due to yaw, would minimise errors. This may have been due to his legal training: to a lawyer, people are always the problem, and the more constrained they are, the fewer difficulties they create. Pollen's mechanistic approach ultimately led him to the synthetic method of fire control. He began with what he imagined would be the ideal rangefinder, a two-man system, feeding an automatic plotter. The Royal Navy never bought Pollen's plotter, but by 1908 it had accepted his idea that plotting was the way to set the Dumaresq. This was not obvious: for example, the contemporary Imperial German navy never adopted plotting.

The bearing rate derived from a bearing plot could, similarly, be applied to a bearing clock, which would project ahead target bearing. In 1908 bearing rate was almost impossible to measure: ships yawed continuously, and gyros were in their infancy. Only in 1911 did the Royal Navy have the means to plot bearings using a gyro-compass. Given a bearing rate, Dreyer recognised that a Dumaresq could be driven backwards. If range rate and speed across (derived from bearing rate and range) and the line of sight to the enemy were known, the pointer for enemy course and the pin for enemy speed could be set. Dreyer called such deduction his 'cross-cut system'.

In May 1908, to support plotting experiments at the year's Battle Practice, the Board of Admiralty approved the purchase of 142 roller boards, squared (graph) paper, and T-squares. This equipment was not ready in time, but fifteen to twenty gunnery lieutenants received manual straight-course (true-course) plotters from Pollen, and others copied that equipment.[9] Of thirty-three ships producing true-course plots, eighteen found them useful. Another eight produced virtual-course plots (four found them useful. Another thirteen plotted range versus time (four found them useful). Attempts to combine virtual course and range versus time apparently failed. This experience was encouraging enough for the Admiralty to buy more equipment for the 1909 Battle Practice, including 121 manual course plotters (with variable-speed paper drive), which automatically plotted own-ship position. Pollen supplied thirteen plotters of his own. Both boards took account of turns by pivoting the board based on the known turning circle of the ship at various speeds and helm angles. All but one ship produced true-course plots. Of thirty-nine ships, eleven measured the target course to within half a point (5⅝ degrees) and sixteen to within one point (11¼ degrees), but average time to obtain this data was five minutes twelve seconds, and accuracy fell badly once firing began. Bearings were too difficult to measure accurately, and there was no way to detect a ship's yaw. Plotting was clearly promising but immature.

Further experiments were depressing. On passage to Vigo in 1910, fleet units tried to plot the courses of all the merchant ships they encountered. They were successful about a third of the time. Reporting tactical experiments, Home Fleet commander Admiral

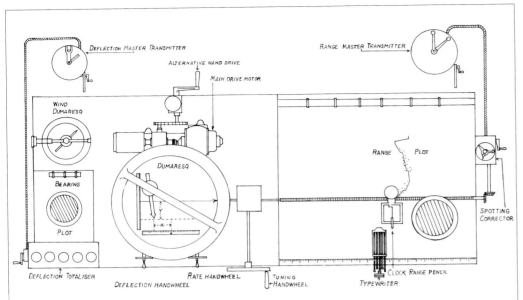

This plan view of a Dreyer Table Mk IV is based on one in the 1916 manual maintained by HMS *Royal Oak*. This in turn was based on a drawing produced by the manufacturer, Elliott Brothers. Note the relative widths of the range and bearing plots; the range plot was the basis of the system. The range plot shows a pencilled clock range line running between scattered points indicating rangefinder readings. The typewriter visible at the bottom was used to enter data from multiple rangefinders, the table operator averaging the data by eye to estimate average range or rate. Deflection was taken from a variety of sources, not limited to the bearing plot, hence the need for a totaliser to sum deflection before it could be sent to the guns. The multiple hand-wheels give some idea of how the table was used: it was constantly 'tuned' to take observed errors into account. Not visible here is the clock drive, which controlled the long screw moving the clock range pencil back and forth to indicate computed (clock) range. The grid visible alongside the pencil carrier was used to measure range rate; a similar grid is visible on the bearing plot. The long screw also drove another shaft providing range to the spotting corrector. This was a differential, adding whatever spots were cranked in to the clock range for transmission via flexible shaft to the range master transmitter, whence ranges were sent to the guns. Ranges were translated to elevation angles at the guns. That was a survival of pre-director practice. The director measured an elevation angle, and it had to be introduced at the guns. This complication was inescapable in a system that had been created organically rather than designed as an integrated unit. The postwar analogue system built around an Admiralty Fire-Control Table was radically different in its basic architecture. Its computer sent elevation *angles* to the guns. Also not visible here are the scribed cylinders used to convert between Dumaresq rates-across and bearing rates.
(A D Baker III)

Dreyer's Original Table, as shown in the 1917 Dreyer Table Handbook.

William H May said that with experience, enemy course could be estimated by eye to within a point, an error equivalent to a range rate of 100 yards per minute. In poor weather it would be folly to hold fire while waiting for a plot to form. In better weather, manoeuvering would make plotting almost impossible. An officer told Pollen that speed estimates by eye were typically 15–30 per cent off. On the other hand, the fleet was successfully using range plots to set the range clock.

Plotting instructions issued in February 1911 warned against virtual-course plotting, because bearings could not be taken accurately enough.[10] The preferred technique was Dreyer's: plot range against time, derive a range rate and set a clock, correcting as necessary when the clock did not agree with observed ranges. Once good gyro-compasses were available, bearing plots would become viable. Even then the Admiralty Gunnery Branch preferred separate range- and bearing-rate plots, which it considered better than an integrated plot (true or virtual motion) in the face of faulty (ie, realistic) data. Rates based on smoothed data would be used to set a Dumaresq by cross-cut, and rates then derived from it. This was a manual version of the Dreyer Table. The next year the Royal Navy formally abandoned true- and virtual-motion plots. Pollen argued that his emerging gyro-stabilised true-motion plotter could overcome the usual errors; it was part of the fire-control system he was developing.

With a reliable gyro-compass, a few years later the US Navy later found manual true-course plotting (which it called tracking) quite practicable. Tracking boards were standard in US warships throughout World War II. They provided both an input into fire-control computers (enemy course and speed) and a fall-back in the event that the computers failed. It may have been significant that the US Navy tended to operate in calmer waters, where yawing would be less of a problem.

The Dreyer Table

By 1909 Dreyer proposed linking his range-time plot to a range clock. At sea that year he built a prototype plotter that proved effective in battle practice. He submitted it to the Ordnance Department in 1910 and patented it that year. Initially Dreyer proposed his table for local turret control; he apparently assumed that Pollen's clock, described below (see pages 53–63), would be installed as the primary fire-control device in a ship's transmitting station. However, he soon saw his table as a viable alternative to Pollen's system. He turned to a precision manufacturer, Elliott Bros (which made Dumaresqs) to turn it into production hardware; his chief designer was Keith Elphinstone. This combination of Dumaresq, plotters, range clock, and transmitter on a single frame was the Dreyer Table.[11] A prototype was tested on board HMS *Prince of Wales* late in 1911.[12] Elphinstone was already working on an improved version incorporating a clock designed for tuning while running (the Vickers Clock was not intended for such tuning, although clearly that was often done).

Dreyer seems to have considered his prototype 'Mk I', although that designation was never assigned. In the 1918 official handbook, it is the device designated the 'original table'. For the planned 1912 comparative trials, Pollen's clock (plus Dreyer's plots) became the Mk II table, although initially Dreyer considered his first seagoing prototype Mk II. Dreyer's improved Mk III table was tested on board the superdreadnought HMS *Monarch* in 1912.

In 1912 the Admiralty decided to adopt the Dreyer Table rather than the Pollen Clock (see pages 53–63) as its fire-control calculator. The production version was Mk IV.[13] The first five were on order by July 1914 (for the *Iron Duke* class and HMS *Tiger*). At the time of Jutland – of the dreadnoughts – only HMS *Agincourt* may not yet have been fitted with a table.[14] During World War I the Mk I and Mk III designations were re-used for simplified versions of Mk IV.[15] There was also a turret table (without any bearing plot), which received no Mark number, but which was widely used. It seems to have been under development by the first half of 1914, and was the local control table Dreyer originally proposed. Cruiser installations began in 1916 with the *Raleigh* (*Frobisher*) class, the simplified Mk I* and Mk III* being ordered for *C*-, *D*-, and *E*-class cruisers. A few light cruisers had turret tables rather than full Dreyer Tables in their transmitting stations. Dreyer Tables survived postwar on board unmodernised capital ships as well as light cruisers and monitors. The last surviving example, a Mk III table aboard the monitor *Roberts*, was not discarded until 1965.

As Dreyer began work, the Royal Navy was buying Pollen's (Argo) stabilised rangefinder mounting, which transmitted range via a step-by-step transmitter. Dreyer saw it as a way of automatically inserting range data onto his plot, and the new stabilised mounting as a way of gaining accurate bearing data. Automation made for much more frequent observations, so plotting became far more practical (in 1910, just before this automation was introduced, the general feeling in the fleet was that plotting was a dead end). In the original Dreyer Table, the step-by-step range transmitter moved a pin across the moving paper. It pricked a hole in the paper indicating the rangefinder reading (bearing transmission was more complex). Pollen's use of the rangefinder was less automated: it lit lamps indicating the range.

The Dreyer Table reflected Dreyer's cautious attitude towards data. As in the pre-Table system, the Dumaresq continued to be central. In Dreyer's view, it and the range plot were two alternative sources of range rate for the range clock. For example, the Dumaresq could be used to give an initial estimate, which could be corrected once reasonably accurate rangefinder ranges could be obtained. If rangefinding became difficult, but enemy bearing could still be measured, a Dumaresq set properly for enemy course and speed (by cross-cut) would produce reliable range rates. The original table had a knob by means of which an operator could adjust the Dumaresq for bearing rate (presumably taking range into account). As in past practice, the Dumaresq helped the operator evaluate a given rate estimate, given the 'bird's-eye view' it presented.

The Dreyer Table carried range and bearing plots, above each of which was a gridded dial. An operator turning the dial to parallel the apparent trend of the plot measured its slope. He called out the rate (range or bearing). In Dreyer's original design, although there was a bearing plot, there was only a range clock, not a bearing clock. The clock was set by a manual follow-up from the Dumaresq, not from the range plot. The clock turned a screw (whose position indicated range) that moved a pencil across the plotting paper. The pencil thus drew a line on the plot indicating clock range. This feedback made it obvious when the clock rate was wrong, and the clock had to be reset. Through a differential, in which corrections (eg, for time of flight and for spotting errors) were added, the range screw drove a gun-range counter. It also drove a master transmitter via a bicycle chain, its reading sent to the guns (or, later, the director) via a follow-up.[16] To avoid possible error, the Dumaresq did not automatically reset the range clock; the officer in charge had to approve setting a new range rate. Dreyer emphasised that the device was advisory; the final decision on rates lay with the gunnery officer aloft, observing the situation. The gun-range transmitter was a late addition, as initially a sliding scale for gun range was mounted on the range screw, with a red pencil to show gun versus present range. Another feature was a means of resetting the clock range.

The 1913 Mk III version introduced a bearing clock, and had the own-ship bar of the Dumaresq geared to a gyrocompass repeater. Thus it would automatically compensate for a change of course by the firing ship. Because the range rate was applied by a follow-the-pointer follow-up, the operator could delay changing the rate to allow for the ship's advance (on her original course) into the turn.[17] The obvious physical difference between the two tables was that in the 1911 version the Dumaresq and range clock were on one side, with the plots side by side, whereas in Mk III the clock (now for bearing as well as range) was placed between the two plots. The feedback for bearing was that the bearing clock output drove the bearing ring of the Dumaresq (there was no bearing feedback pencil).[18] Bearing rate was not the same as the rate across on the Dumaresq, because the relationship varied with range. Deflection was taken not from the bearing plot but rather from the Dumaresq. The 1913 version had a drum with engraved lines to convert bearing rate to rate across for ranges of 2000 to 16,400 yards. The drum was driven by the range clock, so was set for current range. Using the drum, an operator could convert the bearing rate from the plot into a rate across which could be set on the Dumaresq; conversely, the drum could translate the rate across given by the Dumaresq into a bearing rate for comparison with the rate from the plot or the bearing clock.[19] The production Mk IV had two deflection drums, one to translate between rate across and bearing rate (at a given range), the other to give gun deflection at a given range (ie, rate multiplied by time of flight). Three of the latter were provided, for full and reduced charges, and for sub-calibre firing.

The rates shown by the Dumaresq were indicated by the position of the line-of-sight bar on the engraved disc, not by pointers giving precise figures. Range rate lines were 100 yards/minute apart, and rate-across lines were four knots apart. Interpolation limited accuracy, particularly under the stress of battle. The range-rate operator in effect matched a pointer, so his readings were probably accurate to within twenty-five yards/minute. However, the rate across had to be set manually into the bearing clock. It was read off the Dumaresq (probably accurate to within two knots), then translated using the revolving range drum, which added further errors.[20] This limitation was overcome with the 'electrical Dumaresq' of the Mk IV Table.[21]

Dreyer's proposals for Mk IV envisaged plotting data from multiple rangefinders automatically. After a pneumatic device was proposed, in May 1914 the prototype of a typewriter invented by Commander J Brownrigg was ordered (the pneumatic device was formally dropped in July). Nine more typewriters were ordered in August 1914. This became the sole rangefinder plotter in Mk IV and IV* Tables.[22] It was manual rather than automatic. The operator averaged ranges by eye to decide the range to use. Such graphic range averaging was an important feature of later British fire-control systems, and was also used by other navies.

In 1914 rangefinder averaging was key to an emerging concept of rangefinder control tailored to medium-range combat (probably 10,000-yard range

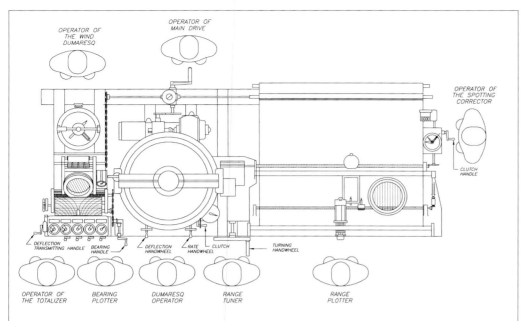

Operators of a standard late-war capital-ship Dreyer Table are shown here. Tuning meant finding corrections that would make the table's predictions correspond with observed reality. By 1918, however, it was fairly clear that prediction would be difficult or impossible, because standard German practice was to zigzag under fire. As from early in World War I, the main virtue of the table was that it presented all available data, so that ranges could be averaged and the best figures chosen. Separate from the table proper was the Dreyer Calculator (developed by Captain Dreyer's Royal Artillery-officer brother), which provided corrections for own and target movement. An earlier version of this is drawing illustrated in William Schlieuhauf, 'The Dumaresq and the Dreyer' Pt. 2, *Warship International* No. 2 (2001). Reproduced by courtesy of the editor and of the illustrator. (W J JURENS)

and below). Salvo firing and spotting were abandoned in favour of individual gunlaying using a precise (averaged) rangefinder range corrected using a few spotting rounds. Range-keeping did not matter in this scheme, as the range would periodically be readjusted according to new measurements. As will be explained in chapter 4, it appears that rangefinder control was emerging as the main British technique at the outbreak of World War I, longer-range methods having proven inadequate. The key was Dreyer's perception that a human operator could rapidly average data by eye, throwing out wild ranges, producing reliable and consistent range data. Plots were no longer mainly a means of finding the range rate.

The Dreyer Table required that the clock be tuned while running. Elphinstone therefore modified the Vickers Clock mechanism.[23] He used a relatively soft iron driving wheel to improve traction and lubricated the disk surface with oil to make it easier to move the roller of the output shaft (to change its speed). The roller was held down by springs to keep it from jumping as it moved. Mk III and later tables had two drive wheels on a common shaft, turning at 15rpm, for separate range and bearing integration. The bearing roller controlled the position of the line-of-sight bar on the Dumaresq. Elphinstone's solution had its problems. The soft driving wheel would wear, particularly if the roller were not moved very frequently. Wear would probably reduce traction on the roller, making for slippage (inaccurate registration of a new rate). Slippage would worsen as the load on the roller increased. The construction of the table, moreover, exposed the driving wheel to dust and grit, which probably tended to wear it down over time.

The Mk V designation was first applied to the improved table installed on board HMS *Ramilles* in 1917, but it was then applied to the redesigned table for HMS *Hood* and her sister ships, installed only on board *Hood* because the others in the class were cancelled.

In August 1915 the range scale was changed from the original 2,000 to 17,000 yards to a maximum of 27,000 yards, as battle range had increased unexpectedly (in HMS *Hood*, the limits were 2500 and 30,000 yards). The table was also modified to allow for the much higher speeds of the new *Repulse* and *Courageous* classes. At long ranges bearing changes were small, so in December 1915 *Queen Elizabeth* proposed that the slope of the bearing line be increased to make it visible. At that time bearing plots were relatively inaccurate, since bearing was being transmitted in quarter-degree steps.

A wind Dumaresq (giving wind speeds along and across corresponding to wind speed and direction) was added in the first half of 1917. This gave contributions to range rate and to deflection based on estimated wind speed and direction (assuming it was constant at the altitudes through which the shell would pass). A totaliser now added the wind across to other corrections for deflection, such as drift (taken from a table for different ranges) and spots. The totaliser used a corrected Dumaresq deflection taken from a second deflection drum, marked to take account of time of flight to various ranges (three alternative drums were provided, for full, reduced, and sub-calibre charges). The deflection read off the corrected drum was

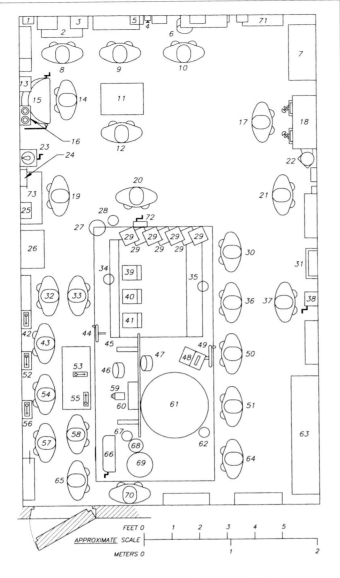

By the middle of the interwar period the Dreyer Table had become considerably more complicated; this 1930s table has ten operators, compared to eight in 1918. Note the separate Dreyer Calculator (73) and the position-in-line instrument (15) needed for concentration fire using a master ship. The ship has an inclinometer, whose readings are received at the table (the receiver is 48), to be used to help tune it. Number 55 indicates the Morse key for aircraft radio (W/T), so presumably the table on which it stands is used to register aircraft spotting data. The new-generation automated tables had special aircraft spotting dials to plot such data, but this installation lacks any such feature. An important new feature, compared to the situation in 1918, was Gyro Director Training (GDT), which provided a limited degree of capability if the target was obscured; its operator is 64. He is one of the two additional personnel serving the table itself, the other being the aircraft W/T operator (58). As before, the main features of the table are its Dumaresq and its bearing plot; the latter carries three feedback indicators: a gun range repeater (39), a range receiver from the spotting top (40), and a clock (datum) range repeater (41). An earlier version of this drawing is illustrated in William Schlieuhauf, 'The Dumaresq and the Dreyer' Pt. 2, *Warship International* No. 2 (2001). Reproduced by courtesy of the editor and of the illustrator.
(W J JURENS)

1 Cease Fire Bell
2 Master Range Transmitter
3 Gun Ready Lamps
4 Fire Control Gong
5 Vickers' Range & Deflection Transmitter
6 Master Deflection Transmitter
7 Evershed Selector Switch
8 Master Range Transmitter Worker
9 Range Phone & Fire Gong Operator
10 Master Deflection Transmitter Worker
11 Fall of Shot Instrument
12 Fall of Shot Operator
13 Distance of Datum Ship Receiver
14 Position-in-Line Operator
15 Position-in-Line Instrument
16 Adding Dial
17 Exchange Operator
18 Telephone Exchange
19 Dreyer Calculator Operator
20 Range Spotting Corrector Worker
21 Order Phone Operator
22 Order Phones
23 Transmitter for Dreyer Correction
24 FTP Range Repeat Receiver
25 Rate Receiver
26 Director Change Over Switch
27 Electro Megaphone to Spotting Officer
28 Mk IV Phone to Spotting Officer
29 Range Receivers
30 Table Diary, etc
31 Grouping Switch for Electric Megaphones
32 Sig Remote Aldis
33 Table Tuner
34 Phone
35 Mk IV Phone to Rate Officer
36 Range Plotter
37 Mean Range Transmitter Worker
38 Mean Range Transmitter
39 Gun Range Repeat Receiver from Table
40 Range Receiver From Dial Reader in Top
41 Datum Range Repeat Receiver from Table
42 Remote Aldis Key
43 Tel W/T Remote Control
44 Tuning Hand Wheel
45 Position-in-Line Gear
46 Deflection Recover
47 Compass Receiver
48 Inclination Receiver
49 Tuning Hand Wheel
50 Range Reader
51 Dumaresq Operator
52 'P' Buzzer Key
53 [G] W/T Key
54 Telegraphist 'P' Buzzer
55 Aircraft W/T Key
56 'Q' Buzzer Key
57 Tel 'Q' Buzzer
58 Telegraphist Aircraft W/T
59 Alternative Hand Drive
60 Compass Control Gear
61 Table Dumaresq
62 Mk IV Phone to Rate Officer
63 Fire-Control Switchboard
64 GDT Operator
65 Deflection Calculator Operator
66 Deflection Calculator
67 Mk IV Phone To Spotting Officer
68 Electro Megaphone Recover to Spotting
69 Wind Dumaresq
70 Wind Dumaresq Operator
71 MAC Repeat Recovers
72 Spotting Correction Handle
73 Dreyer Calculator

compared with that set on the Dumaresq, the corrected deflection being read by the totaliser operator and set on the totaliser.[24] This and each other element of total deflection was set by hand on a separate shaft, the total being calculated by differential gears and indicated on a pointer. Matching the pointer transmitted total deflection to the deflection master transmitter, where another pointer was matched to transmit deflection to the guns or to the director.

The final wartime development was a table, initially designated Mk V, installed on board HMS *Ramilles* in 1917. It had a new bearing transmitter accurate to within four minutes of arc ($\frac{1}{15}$ degree); this improvement was cancelled for other ships because the new Gyro Director Training (GDT) (see below) embodied this improvement and used the director as a bearing transmitter. In this table the second deflection drum was replaced by a direct link to the totaliser, to eliminate the copying function. The Mk V designation was ultimately applied to a redesigned table for HMS *Hood* and her abortive sister ships.

The other major wartime change was Gyro Director Training (GDT), a means of keeping track of a target despite loss of visibility. It was first tried in the monitor *General Crauford* in 1916, to enable her to fire at shore targets when the firing mark was not visible. The Vice Admiral of the Dover Patrol (in charge of the monitors off the Belgian coast) ordered twenty-four more GDT devices in March 1918, and the idea was found useful in other ships. In the monitor the device was simplified by making the change of target bearing manual. However, it was clearly useful for other ships, so by the end of June 1918 it was on board not only the monitor *Lord Clive* but also the battleship *Emperor of India*. There was particular interest in such a device because German light cruisers had hidden behind a smoke screen during an action on 17 November 1917. The design of the GDT for capital ships and cruisers was largely due to Lieutenant Dove RN and to Lieutenant Henry Clausen RNVR, the latter becoming the principal Royal Navy fire-control designer in the 1920s. GDT combined the director, gyro-compass, and

(continued on page 52)

HMS *Hood* had the ultimate Dreyer Table, Mk V (this designation was used for an improved Mk IV aboard *Royal Oak*, then dropped because *Hood*'s was so much improved). The dashed lines are voice pipes above the operators, tying the table to operators moved away from it to reduce clutter (there were no synchros to follow-up table or remote readings automatically). Redesign was demanded because existing transmitting stations were far too cramped. This one was clearly roomier, with a separate space (60) for the bank of radio (wireless) operators it required. They included an aircraft radio operator (72) and a gunnery operator (73), the latter to support concentration fire. Operator (75) was responsible for another form of communication, a remote-controlled Aldis (blinker) light. The other two operated buzzers for salvo fire. In addition, there was a remote-control gunnery radio operator (19) at the table, with a voice tube to the radio room (23). Next to him was the concentration officer (20). Another means of clearing the table was to move some elements overhead, such as the Evershed bearing receiver (44). It showed the director bearing as a light shining on the Dumaresq (43) below it. Other gunnery functions maintained away from the table proper were position in line (PIL, for concentration fire) (console 10, operator 11), the Dreyer Corrector (operator at 12, using receivers [from the table] of range and bearing rates [13 and 14]). The PIL operator used a receiver from a tactical rangefinder indicating the range to the datum ship from which ranges to the target were measured (17). Own-ship range receivers (5) were separated from the table, with their own plotters for rangefinder data (6 and 7). For concentration fire, the compartment accommodated two range receivers for consort data (4), with their own plotter (3). Their range data had to be processed by the PIL operator, to show the equivalent ranges from *Hood* herself. Another element of group fire was a clock marked in sectors during which the ship could fire (16, operator 15, at top left, with voice pipe [42] to the table, emerging near the GDT operator [45]). Plotters were responsible for entering ranges onto the table, using typewriter transmitters. This separation was repeated in the postwar Admiralty Fire-Control Table, in which data were entered and averaged on dummy displays before being transmitted to the computer. Range data could also be received by phone (operator, 56; he was also responsible for the firing gong) connected by voice pipe (28) to the table and to the deflection operator (18). As in earlier transmitting stations, range data were sent to a master transmitter (operator, 58; transmitter 118). The table itself shows the usual split between bearing and range elements, with the range part at the top. This drawing does not show any plots, but it does show a mean rangefinder-range transmitter (55) with its operator (50) alongside a diary keeper (49) who maintained a record of the shoot; the usual range spotter corrector (38) with its operator (53), and a datum range transmitter (37; receiver, 25) for concentration fire. The deflection repeater (33) was near the Dumaresq, so that the two could be compared. Wind conditions were displayed on the deflection or bearing side of the table (speed, 24, direction, 30). A schematic of the table shows four separate long screws carrying pens to mark the plot: gun range, clock range, own (rangefinder) range, and consort range, with a straddle correction bar which could be moved across the plot to see whether all the ranges were close enough. The consort range screw carried a typewriter, so that several consort ranges could be plotted in parallel. Similarly, there was a typewriter for own range, to take account of the ship's multiple rangefinders. This table was designed to exploit inclinometer data (receiver, 35; inclination instrument, 63). The supervising officer (49) stood on the range side, where he could see the situation summarised by the range plot. Note that the space was also used for torpedo control (operator, 62). The space was flanked by dredger hoists for the ship's 5.5in secondary guns (83). In the elevation view at top, range transmitters to the four turrets are shown (107) near the 15in range repeater from exposed rangefinders (117), the range repeater from the turret rangefinders (116), and the master range transmitter (118). Range and deflection receivers for the 5.5in battery are shown as 110 and 111 (one for each side of the ship). Many of the objects in the plan view are phone jacks (26, 36, 40, 54, 59, 99); the telephone exchange operator is 68. This drawing originally illustrated William Schlieuhauf, 'The Dumaresq and the Dreyer' Pt. 3, *Warship International* No 3 (2001). Reproduced by courtesy of the editor and of the illustrator.
(W J JURENS)

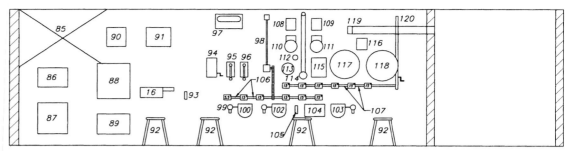

Elevation A

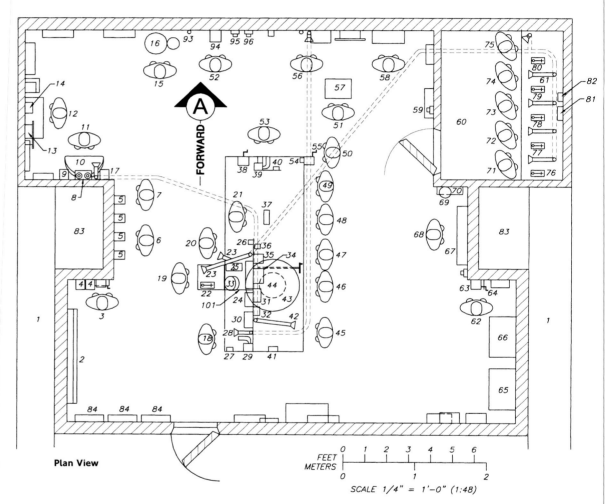

Plan View

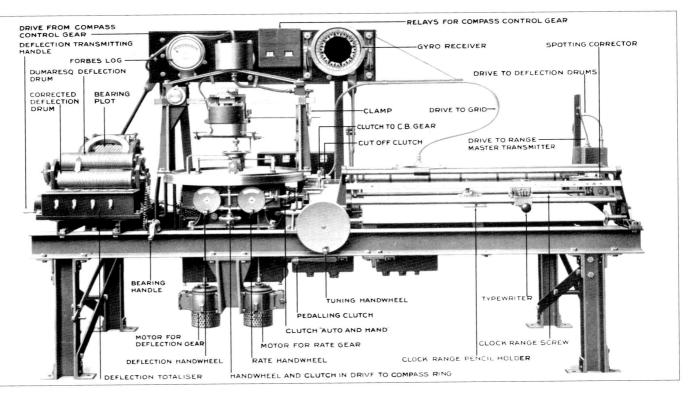

The Dreyer Table Mk IV equipped British capital ships during and long after World War I. Range scale was originally 2,000 to 17,000 yards, but it was extended to 25,000 yards during World War I. Mk IV* began with a maximum range of 20,000 yards, but that was extended to 28,000 yards.
(PROFESSOR JON TETSURO SUMIDA)

The 'Electric Dumaresq' of a Mk IV* Table is shown, in a photograph taken from the 1917 Dreyer Table handbook.
(PROFESSOR JON TETSURO SUMIDA)

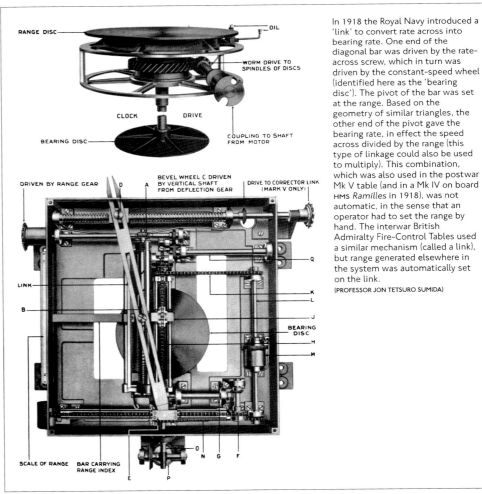

In 1918 the Royal Navy introduced a 'link' to convert rate across into bearing rate. One end of the diagonal bar was driven by the rate-across screw, which in turn was driven by the constant-speed wheel (identified here as the 'bearing disc'). The pivot of the bar was set at the range. Based on the geometry of similar triangles, the other end of the pivot gave the bearing rate, in effect the speed across divided by the range (this type of linkage could also be used to multiply). This combination, which was also used in the postwar Mk V table (and in a Mk IV on board HMS *Ramilles* in 1918), was not automatic, in the sense that an operator had to set the range by hand. The interwar British Admiralty Fire-Control Tables used a similar mechanism (called a link), but range generated elsewhere in the system was automatically set on the link.
(PROFESSOR JON TETSURO SUMIDA)

Dreyer Tables for smaller ships had conventional Dumaresqs without the electrical linkages of the Mk IV version. This is a Mk VI Dumaresq on a Mk III table, as shown in the 1917 *Dreyer Table Handbook*. Its fore-and-aft bar was controlled by the ship's gyro-compass. The operator could translate the rate across given by the Dumaresq into a bearing rate (using the range curves), and set it on the bearing clock using the hand-wheel shown. That was satisfactory only so long as range and bearing rates were low. At high range rates the delay in converting range across to bearing rate would introduce considerable errors even at low bearing rates. The table used electric power for its clock and for its paper drives. Range was 2,000 to 16,400 yards (later extended to 24,400 yards). In June 1918 Mk III tables equipped the *King George V*-class dreadnoughts, and Mk III* equipped C- and D-class cruisers: *Cairo, Calcutta, Capetown, Carlisle, Columbo, Dehli, Dunedin,* and *Durban*. (PROFESSOR JON TETSURO SUMIDA)

(continued from page 49)

the bearing element of the Dreyer Table. It kept the director pointed at the target even when the target was invisible. While the enemy was visible, the director was kept on the target, transmitting its bearing to the receiver in the table. The operator at the table kept track of the series of bearings, eliminating yaw by sight. Comparing the output of a bearing clock with current observation was seen as a very early indication of enemy changes of course. After trials in HMS *Emperor of India*, GDT was approved for service in September 1918, and an order for twenty-four more sets was approved in January 1919 (six had been completed by December).

Associated with GDT was a new kind of straight-line plot. It showed only deviations from the bearing-clock reading, so that it was easier to see whether the target, when visible, was still at the expected bearing. This type of plot, applied to both range and bearing, was adopted for the postwar Admiralty Fire-Control Table. It was first tried in HMS *Queen Elizabeth* and in HMS *Renown*. Own-ship movement was eliminated from the plot. Errors in rate and enemy manoeuvres

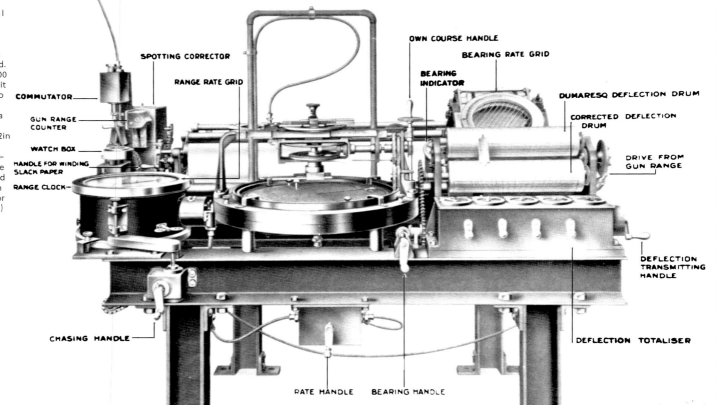

The Dreyer Table Mk I was the simplest of the series retaining range and bearing plots. The table was entirely hand-worked. Range scale was 2,000 to 20,000 yards, but it could be extended to 28,000 yards. The clock was driven by a spring. In June 1918 Mk I equipped the 12in dreadnoughts. Mk I* equpped the *Raleigh*-class cruisers and the monitors *Glatton* and *Gorgon*. (Photograph courtesy of Professor Jon Tetsuro Sumida.)

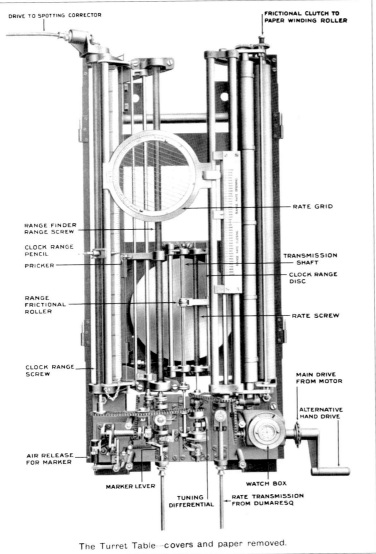

The range-only turret table was designed for battleship turrets, as a means of secondary control, but it also equipped some cruisers. During World War I these tables were given to the Royal Italian Navy to equip battleship transmitting stations (they were superseded by Le Prieur systems). In this photograph from the 1917 *Dreyer Table Handbook*, paper and covers have been removed. In June 1918 turret tables equipped the light cruisers *Ceres*, *Chester*, *Carysfort*, *Comus*, *Melbourne*, and *Royalist*. By about 1930 any cruiser which did not have a Mk III* had a turret table: *Calliope*, *Centaur*, *Cardiff*, *Cambrian*, *Concord*, *Canterbury*, *Coventry*, *Castor*, *Caledon*, *Curacao*, *Champion*, *Calypso*, *Curlew*, *Constance*, *Carysfort*, *Cleopatra*, *Comus*, *Conquest*, *Brisbane*, and *Dartmouth*.
(PROFESSOR JON TETSURO SUMIDA)

by a consort firing at the same target, but it was not standardised in wartime (there was some fear that the plot would become invisible if too many carriages and pencils were added). The Mk V table in HMS *Hood* was redesigned so that the typewriter ranges, consort range (for up to three consorts), and fall of shot could be applied remotely, reducing clutter at the table. This remote-entry idea was incorporated in the postwar Admiralty Fire-Control Table.

The wartime Mk I Table was hand-worked, with paper run by a spring-drive clock. The range plot was calibrated for 2000 to 20,000 yards on a scale of 400 yards to one in speed. The extended scale for the typewriter was graduated up to 28,000 yards. Mk I* had a gyro-controlled Dumaresq own-ship bar and a master transmitter for range. Mk III was midway between the full Mk IV and the stripped-down Mk I. Apart from an electrical main drive to the clock discs and paper rollers, it was hand-worked (Mks IV and V were fully electrical). The extended range scale was graduated to 24,400 yards. This version had a master transmitter to the guns. Mk III* had an automatic corrector link transmitting corrected deflection to a totaliser for transmission to the guns.

The full Dreyer Table was apparently never released to other navies. However, the simpler turret table (range only) seems to have been provided to the Royal Italian navy in wartime, and it was probably either given to or at least shown to the Japanese.

In 1917 an experienced British gunnery officer, Commander R T Down RN, visited Washington and described the British plotting system in detail. The US reaction was incredulous. After creating an elaborate rate plot, the British relied on the officer in the top with his Dumaresq; he was authorised to decide what rate to use. To an American fire-control developer, that meant that 'after all their elaborate system, they show that they are afraid of it and fall back on simpler methods. In the Battle of Jutland, they simply set the rate at zero, because they had no idea of what it really was.'[25]

Pollen and the synthetic solution

If Dreyer's was the ultimate analytic solution to fire control, Pollen's was the first synthetic one. His back-

were instantly visible. The plotting paper could be quite narrow, making it possible for the plotting officer to remain in one place whatever the change in range. In *Queen Elizabeth* that made space for a radio operator to sit at the table alongside the plotting officer. Using a narrow plot with a moving range scale made it possible to plot to great ranges without shrinking the scale, and to exploit information from aircraft.

During the war ships increasingly saw the table as a summary plot rather than as a way of predicting range. After experiments in various ships, in August 1917 a gun-range pencil was added. Now observations of the fall of shot could be made on the plot, and range corrected accordingly. As concentration tactics developed, some ships added a pencil to show the range reported

ground as a lawyer may have helped him see the gunnery problem as a whole. In effect, Pollen was to fire control what John Holland was to the submarine. Because he was an outsider, he was unaware of the existing system, hence not tempted to try to perfect it. He did not always recognise the key advantages of his system. Like many other inventors he often emphasised what turned out to be minor features, such as an automated true-motion plot.

Pollen's integrated vision convinced many naval officers that he was offering something very important. They saw him as part of an accelerating technological revolution in which they were very active participants, other prominent elements being radio, high-powered guns, and the submarine. Several of these technologies offered potentially high pay-offs, but they were quite risky. Officers were painfully aware that new equipment rarely lived up to its advertising. The more complex the equipment, the less sailor-proof it was. Experienced officers had spent entire careers learning that people often had to compensate for gaps in technology. They considered complexity acceptable only if the simpler equipment could not perform adequately – or if it promised some revolutionary leap in capability. This point of view was not limited to the Royal Navy. The forward-looking contemporary US Navy abandoned power loading for turret guns because man-handling seemed more reliable.

Pollen could not have pressed ahead had he realised how risky his project was. He committed the usual sins of developers of new technology: over-optimism, and a belief that prototype systems were ready for production. These sins were unfamiliar in a world in which development effort tended to concentrate on elements of a system but not on the system as a whole.

The evolving Royal Navy fire-control system put the control officer and his judgement at its centre. Pollen sought the opposite: to remove human judgement as completely as possible. Despite their inversion of the evolving system of fire control, his initial ideas may have been attractive because they promised to eliminate dead time. A more automated ship could function at higher or more variable range rates. If it could impose such rates on a less sophisticated enemy, then it could hit without risking much damage. This much was probably obvious to those who had been following gunnery development since the 1904 trials. Pollen did not emphasise this possibility; he did not yet realise just what he was selling.

The Royal Navy allowed Pollen to test his two-man rangefinder and a manual plotter on board HMS *Jupiter* between November 1905 and January 1906. Like past two-man devices, the rangefinder failed. The manual plotter also failed, because it was slow and prone to error. Pollen concluded that his ideas were good but that the equipment needed further development.

Remarkably, DNO (Captain John Jellicoe) agreed. Pollen showed enough promise to be given what amounted to a system-development contract. He envisaged an integrated system, which he called AC (aim correction) as distinct from fire control (tactical handling). Its 'change-of-range machine' (ie, computer) would correct for yaw (using gyros) and would automatically transmit gun orders. Pollen adopted a gyro-stabilised Barr & Stroud rangefinder but retained the automatic plotting table. The idea of computing based on an

The Argo Mk IV Clock (computer) was offered for export after the Royal Navy withdrew from its agreement with Pollen. The clock mechanism is in the box; the large dial indicates target range. The face of the clock was a Dumaresq-style presentation of target course and speed. Argo included these photographs with the brochures describing its clock and the associated true-course plotter (the brochure and photographs were collected by the US naval attaché in London). Argo also produced a true-course plotter to provide input to the clock.

next commander of HMS *Excellent*, the gunnery school) suggested that the system be modified to accommodate manoeuvres by the firing ship, so that a ship could evade torpedo attacks without having to cease firing.

British gunnery Lieutenant Reginald Plunkett (later Admiral Drax) summarised what was required – and what Pollen was trying to do – in a 1911 paper on fire control.[32] It was widely understood by this time that rates were key to fire control, and that rates changed over time. The Dumaresq gave only a snapshot of the tactical situation at one point in time. Even if own ship and target pursued steady courses at steady speeds, the range rate changed as the line of sight between the two ships moved (the rate at which it changed depended on the speed across). The Dumaresq could be kept updated only if the target was continuously visible. If,

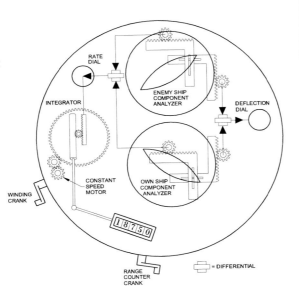

Alongside the main battery Mk I, Ford developed the 'Baby Ford' (Range-keeper Mk II), suited to battleship secondary batteries and to destroyers. A computer for such secondary functions was, if anything, even more revolutionary than one for main batteries. This drawing shows the face of the Baby Ford and the mechanism behind it. There was only a single integrator, for range rate; there was no attempt to convert speed across (deflection) into bearing rate. Own- and enemy-ship speeds across were added and integrated (at left) to provide an estimate of current range. Speeds across were added to give total deflection. Enemy bearing was set by eye. By the 1930s Mk I was being superseded by a full range-keeper, Mk VII, in battleship secondary batteries; it was also used as a fallback computer for cruiser directors. New-generation destroyers had more sophisticated dual-purpose range-keepers integrated into their Mk 33 directors (those with single-purpose main batteries had Mk 35s, with their own new range-keepers). This drawing originally illustrated 'Questions on the Effectiveness of US Navy Battleship Gunnery: Notes on the Origins of US Navy Gun Fire-Control System Range-Keepers,' Pt 3, by C C Wright, *Warship International*, Vol 42, No. 1 (W J JURENS).

however, the line of sight could be automatically updated, a Dumaresq would continuously give the correct (changing) rate. This rate could be fed into an integrator. Plunkett also observed that because both enemy course and speed had to be measured, feedback required that two separate variables be tracked, most likely range and bearing.[33]

Plunkett's description applied to a whole generation of synthetic fire-control computers. A mechanical model of the engagement could be used to project ahead target position and line of sight, hence the rates. For simplicity, Pollen used an equivalent to a virtual plot. The main change after Pollen was that designers separated own and target computation, assigning each its own component of the two rates, and then adding the two for system output. The summed range rates were integrated to give total range; rates across were added, then divided by range to give a total bearing rate, which was integrated to give bearing. This separation made for easier mechanical design, and for easier isolation of own- from target-ship manoeuvres, but it was not a major conceptual leap. In such a configuration, a linkage between the two drives represented the line of sight between own ship and target. Because the system handled enemy course and speed separately from own course and speed, it was inherently independent of own-ship manoeuvres. That is why the US Navy said that, for such a system, tactics drive fire control, ie, the system does not limit own-ship manoeuvres.

Synthetic systems could continue to track (and to engage) a target even when it was no longer visible. Radar largely eliminated that problem, but synthetic operation was still important because it alone could fully predict target motion, hence make hitting possible at long range. An analytic system, no matter how well it smoothes input data, makes linear predictions (constant rates), approximations that become less and less valid at longer ranges. Thus post-World War II radar anti-aircraft systems divided into medium-range linear ones like the US Mk 56 and longer-range synthetic ones like the US Mk 68.

The Mk I Pollen clock tested on board *Natal* was initialised with target range and bearing and with range and bearing rates. A Dumaresq equivalent used what amounted to cross-cuts to deduce target course and speed, feeding them into a virtual plot in analogue form. The special feature of the system was that the plot was mechanically driven, so that once given target course and speed it generated present target range and bearing – it was a synthetic device. Shell time of flight was used to calculate deflection (bearing rate multiplied by time of flight). Pollen used a virtual course plot internally for simplicity. Given own ship speed (an input), it was simple to translate back and forth between true-course and virtual-course data.

The Mk I system on board HMS *Natal* was tested in May and June 1910, her automatic plots being compared with manual plots of true and virtual courses by the battleships *Lord Nelson* and *Africa*. All of the equipment was rejected as too unreliable and too delicate; the devices needed 'more skilled attention than they can get in a newly commissioned ship.' Even so, the

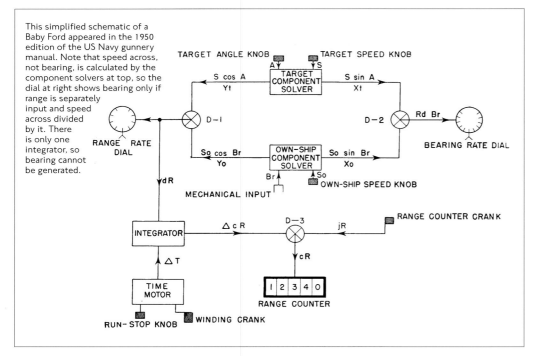

This simplified schematic of a Baby Ford appeared in the 1950 edition of the US Navy gunnery manual. Note that speed across, not bearing, is calculated by the component solvers at top, so the dial at right shows bearing only if range is separately input and speed across divided by it. There is only one integrator, so bearing cannot be generated.

trials committee recommended adopting, with modifications, the stabilised rangefinder (which was done) and the Pollen Clock. The Pollen plotter should be tried again 'when further developed.' Pollen insisted that his plotter (on which he had expended considerable time) was as important as his clock (computer). That made sense in the context of Mk I, which needed a great deal of input, including rates. However, Pollen was already working on a radically different device, a true synthetic system (Mk II). Pollen's insistence on true-course plotting alienated many in the Admiralty who were all too aware of the failure of attempts at true-course plotting.

In 1909 work began on a more flexible Mk II, offering the helm-free operation that Ogilvie later wanted.

Its architecture was completely rethought to create a fully synthetic machine (it is not clear to what extent Pollen realised how radical the change was). The only target inputs were estimated speed and course, taken from a true-course plotter. Other inputs were firing-ship speed and target bearing. From these, virtual course and speed could be computed. A Dumaresq equivalent calculated range rate and rate across. Mk II incorporated Pollen's new trademark 'slipless' integrator (patented in 1911), a ball in contact with the constant-speed disc, driving two parallel rollers in a frame above the disk. Projectile time of flight was used to set gun range (adding to present range by multiplying range rate by time of flight).[34] The Admiralty rejected gun-range correction due to time of flight as an unnecessary complication. Presumably the Dreyer Corrector seemed sufficient. It is not clear to what extent its limitation to low range rates was understood. Note that Mk II did not generate target bearing (ie, bearing rate) because its Dumaresq-equivalent calculated speed-across rather than bearing rate.

Ironically, Mk II did not really need the true-course plot as input. It could just as well have started with measured range and bearing rates (and ranges, to convert the latter into rates across), usable for cross-cuts from which target course and speed could have been deduced. Many later synthetic systems (and the Admiralty version of Pollen's clock) used those inputs, because true-course plotting often proved awkward (the US Navy seems to have preferred true-course plots). True-course plots were later used by many navies to provide their gunners with a useful overall view of the engagement, and they were used as tactical plots.

Once the Admiralty had rejected Pollen's plotting table, he could be seen as no more than the supplier of a system component, his clock. The Dreyer Table was an alternative. For 1912 DNO planned comparative trials.[35] Five sets of Dreyer equipment were bought for the *Orion* class (except HMS *Orion*) and the *Lion* class; and five sets of Argo Clocks with automatic range-time plotters (as in the Dreyer Table), four of them for the *King George V* class and one for the battlecruiser *Queen*

Mary. The requirements of the Argo system explain why the *King George V* class were given much lighter masts with small foretops. The prototype Argo Mk IV Clock was tested on board HMS *Orion* in November 1912.[36] At about the same time a prototype Dreyer Table was tested on board HMS *Monarch*, though under different conditions. HMS *Thunderer* of the *Orion* class may have had an Argo Clock rather than a Dreyer Table for the 1912 director firing trials.

DNO's assistant Commander J C W Henley asked that the clock generate bearings and also that it be modified to use range and bearing rates as alternative inputs. The result was Mk III. The solution, which was significant for later fire control, was automatically to divide rate across by range to generate bearing rate. Conversely, Mk III could multiply an input bearing rate by range to produce a rate across, which could be used for a cross-cut. Mk IV, the final Royal Navy version, was further modified to accept changes in the speed and bearing dials once running. Mk V was a later commercial version sold to the Imperial Russian navy. It incorporated a gyro to correct its bearing setting when the ship turned, and it calculated gun range. In effect it was equivalent to all later fire-control computers, except that it did not handle own and target data separately, hence it had to generate what amounted to a virtual-course plot internally.[37]

Both the Dreyer and Pollen systems ran trials, but never competitively. The Royal Navy bought only the six Pollen systems intended for trials; the Dreyer Table became the standard World War I fire-control system. Although a Dreyer Table required significantly more manpower than a Pollen Clock, it was still far more automated – and faster – than the manual system it replaced. Compared to a Pollen Clock, the Dreyer Table lacked the ability to handle high and varying range rates, and to project target position when the target was obscured, as in typical North Sea mists. These limitations may well not have been understood by the principal decision-makers. Moreover, by 1914 the Dreyer Table fitted much better than the Pollen Clock into an emerging system of rangefinder control suited to new tactics.

Why did Pollen fall from being the Admiralty's great hope for the future, to being an annoying civilian contractor? Because none of the relevant Admiralty policy papers has survived, what follows is somewhat speculative.[38] Beginning about 1907, British Liberal politicians questioned the value of any monopoly agreement like the one the Admiralty was about to sign with Pollen. They argued that free competition would always produce better products at more reasonable prices. The printed series of DNO notes to successors mentions again and again a desire to eliminate the sort of monopoly-secrecy contract Pollen had. This policy was pursued even if the monopoly product was superior. Thus the Royal Navy dropped Vickers' monopoly on British submarines, despite the company's evident technical superiority (the Admiralty bought several admittedly worthless foreign designs).[39] Monopoly could be justified only if a product was unique; that Dreyer had devised a viable alternative suggested that Pollen's was not.

In 1910 Barr & Stroud, until then a supplier of rangefinders and data transmitters, began work on its own fire-control computer system, built around a Dumaresq equivalent, the Rate of Change of Range and Deflection indicator (ROCORD). In 1911 the company hired retired Dutch Rear Admiral W A Mouton specifically to develop a fire-control system.[40] This initiative probably reflected Admiralty encouragement. It was another potential alternative to Pollen's system. ROCORD was supplied in small numbers to the wartime Royal Navy.

In 1912 Pollen's monopoly and secrecy agreements came up for renewal. DNO stated in his hand-over notes for his successor that he wanted to eliminate secrecy agreements altogether. Probably he expected further manufacturers to enter the fire-control field. Even if Pollen's system were better, he would feel compelled to cut his price.[41] Pollen's reaction may have been unexpected. He stood his ground on prices. Unlike Barr & Stroud, he had no other line of products sales of which could absorb his development costs.[42] After successful trials he had begun to tool up for the expected quantity production.

Pollen's system also inverted previous gunnery practice. It concentrated control in a machine in the transmitting station instead of with an officer aloft. Officers who had seen many well-designed machines fail the test of 'sailor-proofing' knew that an officer could function under very difficult conditions. At best the machine might be far better, but at worst the machine – and the ship's main battery – would fail altogether. Pollen left no fall-back path to earlier manual techniques. Thus the argument about reliability may have been crucial.

Pollen, who was still well connected, thought that the 1912 decision reflected a larger Admiralty choice to de-emphasise long-range gunnery. He decided to change his argument for his system, from long-range accuracy to the ability to keep hitting despite radical manoeuvres (the system offered both virtues). The reasoning described in chapter 4 supports the claim of a shift, but the economic argument (and the existence of a less expensive alternative) was probably enough.

Pollen clearly felt ill-used. He had been encouraged to continue despite the disappointment of 1907–8. Now he blamed officers too naive to understand just how revolutionary his system was. The Admiralty's confusing public comments give the impression of squirming to avoid any sort of direct explanation (not an unusual behaviour for a government department). Official statements, though strictly correct, were easily misinterpreted. The Admiralty briefed First Lord Winston Churchill to tell parliament that Pollen's system had been dropped in favour of a much superior navy system (the Dreyer Table). The obscured record may help explain a later comment by Hugh Clausen, the main interwar official British fire-control designer, that it was 'too dangerous' to attempt an official history of British naval fire control.

Although it had been developed in secret, the Pollen system was widely known within the Royal Navy, and it was the focus of great hopes. Many naval officers had invested in Pollen's Argo company. Many in the know wrote and published letters (eg, to *The Times*) doubting the Admiralty's claim. According to the US naval attaché, 'many, if not most, of the officers actually engaged in fire-control work are heartily in favor of the Pollen syste.'. It seems unlikely that he was quoting Pollen, as he also commented that Pollen's resentment of Admiralty action may have coloured his remarks. After the Admiralty dropped him, Pollen 'challenged [it] to test any system in competition with the AC [Argo Clock], but the Admiralty will not allow it.' He

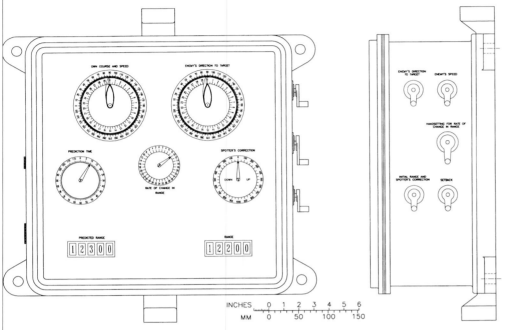

The US Navy Bureau of Ordnance asked Sperry to produce a computer equivalent to the one Pollen was selling. His senior designer, Ford, having quit, Sperry competed with Ford to produce what the bureau wanted. This is Sperry's design. Note that it does not show the own- and enemy-ship picture typical of later synthetic systems. Sperry lost the competition, at least partly because his computer was not ready in time. This drawing originally illustrated 'Questions on the Effectiveness of US Navy Battleship Gunnery: Notes on the Origins of US Navy Gun Fire-Control System Range-Keepers,' Pt 2, by C C Wright, *Warship International*, Vol 41, No 3. (W J JURENS).

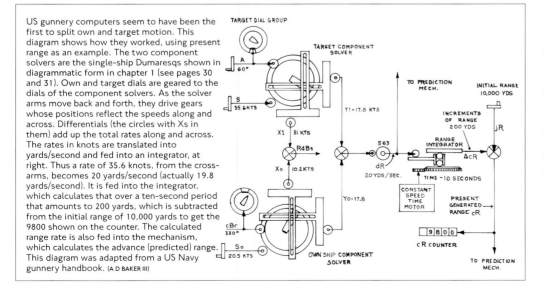

US gunnery computers seem to have been the first to split own and target motion. This diagram shows how they worked, using present range as an example. The two component solvers are the single-ship Dumaresqs shown in diagrammatic form in chapter 1 (see pages 30 and 31). Own and target dials are geared to the dials of the component solvers. As the solver arms move back and forth, they drive gears whose positions reflect the speeds along and across. Differentials (the circles with Xs in them) add up the total rates along and across. The rates in knots are translated into yards/second and fed into an integrator, at right. Thus a rate of 35.6 knots, from the cross-arms, becomes 20 yards/second (actually 19.8 yards/second). It is fed into the integrator, which calculates that over a ten-second period that amounts to 200 yards, which is subtracted from the initial range of 10,000 yards to get the 9800 shown on the counter. The calculated range rate is also fed into the mechanism, which calculates the advance (predicted) range. This diagram was adapted from a US Navy gunnery handbook. (A D BAKER III)

read the Admiralty's inaction as evidence of fear that the AC would show its superiority. However, it is also possible that the Admiralty did not want to show, by the conditions of a comparative trial, that it was now much more interested in shorter-range combat emphasising a high volume of fire (as seems to have been the case; see chapter 4).

During World War I Pollen often pointed to examples of very high range rates in combat, and also to cases in which targets had been obscured – in which analytic systems could not keep shooting. It must therefore be asked whether the Admiralty decision against Pollen in 1912 led directly to disappointing British gunnery performance during World War I.[43] The great wartime surprise was that the Germans, who did not have a system particularly adapted to high rates, adopted high-rate tactics (zigzagging under fire). Pollen's system was designed to enable a *British* fleet to manoeuvre radically while firing at a relative docile enemy (its adoption might have encouraged the British to develop zigzagging tactics which would have frustrated German fire control). Without any Pollen-like system, the Germans did not do very well, either. Both sides had assumed that many hits would be needed to destroy a modern capital ship. The great surprise was that *single hits* destroyed three British battlecruisers. The evidence (see chapter 4) is that extremely dangerous British magazine practices were at fault. They are very indirectly traceable to a belief that long-range fire control would be poor, but that is a long and indirect route to the magazine explosions.

The export market

Once the Admiralty rejected his system, Pollen demanded the right to market it. That was exactly what Vickers received when its monopoly on British submarine construction was cancelled at about this time. However, the Admiralty knew that Pollen's fire-control system was sufficiently in advance of anyone else's to render exports potentially dangerous. It tried to block him by claiming that he was merely exploiting Admiralty research, and that allowing him to go public would be to give away important secrets. The British press published numerous letters from officers arguing that by dropping Pollen's system and by ending the secrecy agreement, the Admiralty had compromised years of gunnery development. The Admiralty threatened Pollen with the Official Secrets Act, but had to

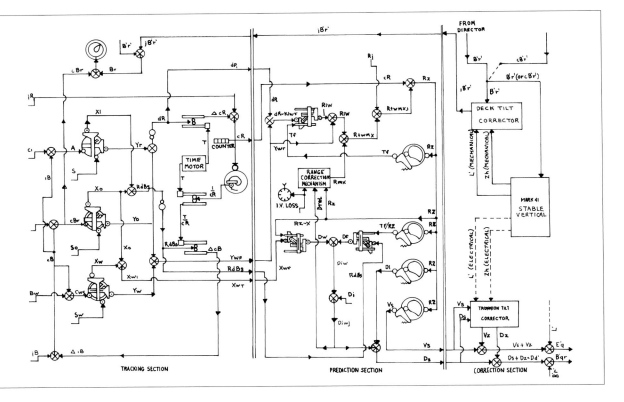

The range-keeper used mechanisms like those illustrated to produce the necessary predictions. This diagram is based on US World War II practice, the system being divided into a tracking section which provided current range and bearing, a prediction section which showed where to aim guns, and a correction section to take into account deck and trunnion tilt, using a stable vertical. The prediction section at left shows three component solvers: one for own ship, one for the target, and one for the wind (suffix w). They feed three integrators, all driven by the same motor. The cam at the right of the tracking section inverts the range (it calculates 1/R) so that range rate across can be turned into bearing rate (the straight device follows the groove in the cam, and the roller reads off its position). In this case, instead of dividing range across, the rate at which the rate-across integrator is driven is set by 1/R. The three push-pull mechanisms in the prediction section are multipliers. The four devices at the right of this section are partial-correction cams. Multiple cams are used so that the size of each cam can be limited. This drawing was adapted from one in a US Navy gunnery manual. Architectures varied. A diagram of Range-keeper Mk VII Mod 5, a widely used auxiliary range-keeper of World War II, shows separate component (target, wind, ship), present range, deflection, advance range, transmitter (to the guns, taking ballistics into account), and time sections.
(A D BAKER III)

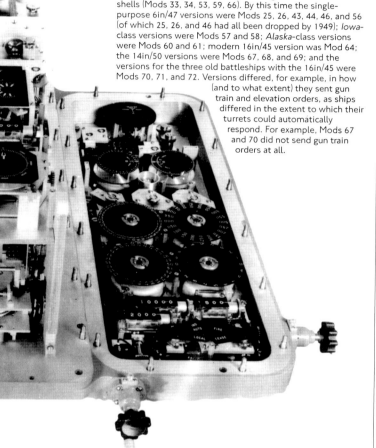

A Range-keeper Mk 8 and its innards, from the 1943 manual for the Mk 38 director. This was the standard main battery range-keeper on board US battleships and cruisers beginning with the rebuilt *New Mexico*-class battleships and the *New Orleans*-class cruisers. Because it was an analogue computer, Mk 8 (and other range-keepers) was built in different versions to control different calibres and even to cater to different sets of ballistics. It also had to be modified to take into account wartime developments in remote control of turret mounts. Mods 0 and 1 were 8in/55 versions for the *Indianapolis* and initial *New Orleans* classes; Mod 2 equipped the rebuilt *New Mexico*s. Mod 3 was another heavy-cruiser version, for the later *New Orleans* class (Mk 31 director). Mod 4 was for the *Brooklyn*-class light cruisers (6in/47 guns, Mk 34 director). Mods 5 and 6 equipped the heavy cruisers *Vincennes* and *Wichita* (Mk 34 directors). Mod 7 was for the *St Louis* class. Mod 8 (1938) was for the *Northampton* class, indicating modernisation of their fire-control systems. There was no outward indication of such modernisation, but Mod 8 was later modified as Mod 30. Modernisation included a separate stable vertical. Mod 9 was for the *North Carolina* class (16in/45 guns). Mod 10, a modified Mod 2, seems to have been planned for gunnery modernisation of the *California* class, but was never placed in service. Mod 11 was for the *Iowa* class. Mod 12 was for 6in/47 main battery (presumably for the *Cleveland* class), and Mod 13 for 8in/55 (presumably *Baltimore*). Mod 14 was for 14in/50 main batteries, probably for the rebuilt *California*s. Mod 15 was for the *Alaska*s (12in/50). Mod 16 was for 16in/45 main-battery guns, probably for the rebuilt *West Virginia*. The sheer variety of wartime upgrades and new versions for existing ballistics suggests the extent of wartime ordnance activity, compared to relatively slow pre-war developments. At least one probably applied to the planned upgrade of the surviving *New Orleans*-class cruisers with Mk 35 directors. Upgrades were Mod 17 (from Mod 7), Mod 18 (from Mod 4), Mod 19 (from Mod 14), Mod 20 (from Mod 0), Mod 21 (from Mod 1), Mod 22 (from Mod 3), Mod 23 (from Mod 5), Mod 24 (from Mod 2), Mod 27 (from Mod 6), Mod 28 (ex Mod 3A), Mod 29 (ex Mod 3B), Mod 30 (from Mod 8). In wartime the series exploded, so that by June 1945 it had reached Mod 63 and by 1949, Mod 72 (enough earlier versions had been dropped by this time that the 1949 handbook applied to only twenty Mods). Note that the June 1945 list does not appear to include the version for the modified *Pennsylvania* (14in/45 ballistics). As of 1949 new applications included *Worcester*-class cruisers with dual-purpose 6in/47 guns (Mod 65) and 8in guns with 335lb AP shells (Mods 33, 34, 53, 59, 66). By this time the single-purpose 6in/47 versions were Mods 25, 26, 43, 44, 46, and 56 (of which 25, 26, and 46 had all been dropped by 1949); *Iowa*-class versions were Mods 57 and 58; *Alaska*-class versions were Mods 60 and 61; modern 16in/45 version was Mod 64; the 14in/50 versions were Mods 67, 68, and 69; and the versions for the three old battleships with the 16in/45 were Mods 70, 71, and 72. Versions differed, for example, in how (and to what extent) they sent gun train and elevation orders, as ships differed in the extent to which their turrets could automatically respond. For example, Mods 67 and 70 did not send gun train orders at all.

relent because Pollen showed that he had proposed his system before he had been exposed to any official gunnery ideas. In effect the Admiralty was admitting that Pollen's ideas were revolutionary. The clearest indication of declassification at the time was the first fire-control article ever published in *Brassey's Naval Annual*, in 1913, which referred to Pollen as the first man to have turned his attention to range and deflection rates.

In October 1913 Pollen distributed a brochure, with photos and drawings.[44] For the first time they described the new kind of fire-control system. The US naval attaché reported that, beside himself, the attachés from Austria, France, Germany, Italy and Russia were most impressed with Pollen's system, and that Austria and France were sending commissions to examine the system.[45] Within a few months Pollen reported an eighteen-month backlog of orders. Before World War I

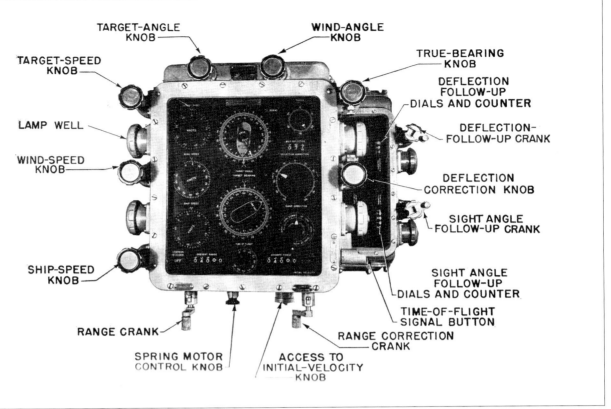

The US Mk VII was a smaller equivalent to the Mk VIII that controlled battleship and cruiser main batteries. The 1944 and 1950 US ordnance texts used it to exemplify modern surface fire-control computing practice. Mk VII controlled battleship single-purpose (5in/51) secondary batteries, and it was also an auxiliary range-keeper in cruisers, where it could be used either for divided fire (ships had one main range-keeper) or in the event the main range-keeper failed. In effect it was a much more capable replacement for the 'Baby Ford'. Unlike that device, it computed advance range and deflection, as in the larger range-keepers. Mk VII corresponded broadly to the British AFCC, although unlike the British unit it was never used as a destroyer computer (the single-purpose destroyer system used Mk XI). Because it had no plotter, Mk VII offered no direct feedback. However, when used to control secondary batteries, its target was usually in direct view. The operator, then, could compare the apparent position of the target-ship outline with the actual apparent course of the target. In 1941 it was being superseded in new and modernised ships by Computer Mk 3. Neither generated bearing (ie, integrated deflection), so neither could be used for blind fire. Mk VII was very widely used; there were seventeen versions, of which Mod 0 controlled 5in/51s aboard *New Mexico*-class battleships and Mods 3 and 5 also controlled 5in/51 guns. Mod 1 was for heavy cruiser after control stations. Mods 2, 4, 8, 11, 14, and 15 were auxiliary 8in computers. Mods 6, 12, and 16 were for 6in/47 batteries. Mod 10 was for the 6in/53 guns of the old *Omaha*-class light cruisers.

broke out he sold two systems to Austria-Hungary and five to Russia. He was close to selling others to Chile (for the *Almirante Latorre* class) and to Brazil (for the projected battleship *Riachuelo*, which was never built). Turkey, Greece, Italy, and France had all shown interest. US interest in buying a test unit and even in license production declined as Sperry and Ford offered to produce equivalents (inspired by Pollen). The outbreak of World War I ended the bulk of the export business, and the Admiralty refused to buy further Pollen fire-control systems (he was dropped formally from the list of approved suppliers). The Austrian systems were 'lost in the mail' at the outbreak of war and were never delivered.[46] The Russian system was delivered, but saw little war service. Although the US Navy did not adopt Pollen's system, its concept almost certainly inspired the development of the Ford-Sperry systems, including the Ford Range-keeper.

Every major navy in the world adopted synthetic systems after World War I, because analytic ones had performed so poorly in combat. Barr & Stroud's rival synthetic system was adopted by the Japanese and the Italians, and probably inspired the German system. Pollen was behind all of the British systems. It is possible that, had the Admiralty paid to suppress Pollen's work altogether, none of these World War II synthetic systems would have appeared. In this rather distant sense, the triumph of *Bismarck*, which had a modern synthetic system, over *Hood*, which had a Dreyer Table, might be characterised as an own goal for the Admiralty. It would have done better to pay Pollen to suppress his system altogether.

As ranges opened, gunners could no longer see shells actually hitting their targets. They had to rely instead on what they could see, the splashes of shells that missed. Given enough shells in a salvo, a few near-misses could indicate hits. This is what a spotter could often see: splashes around a target. They are correct for line (note that *Iowa* has been turning) and they are straddling the target. In this case they were made by USS *Mississippi* firing at the radio-controlled target (ex-battleship) *Iowa*, 22 March 1923. *Iowa* was sunk by gunfire that day.

CHAPTER 3
Shooting and Hitting

THE RUSSO-JAPANESE WAR, which was being fought while the Royal Navy experimented with long-range firing, showed both that heavy guns could actually hit, albeit rarely, at extraordinary ranges and that single hits could be decisive. British naval attaché Captain Pakenham famously remarked that:

> the effect of the fire of every gun is so much less than that of the next larger size, that when 12in guns are firing, shots from 10in pass unnoticed, while, for all the respect they instill, 8in or 6in guns might just as well be pea-shooters...this [refers] entirely to moral effect...Everything in this war has tended to emphasise the vast importance to a ship, at every stage in her career, of carrying some of the heaviest and furthest-shooting guns that can be got into her.

The Japanese now considered 8in and similar guns worthless; they wanted only 12in for their future cruisers and battleships.

In the attaché's account of the Battle of the Yellow Sea, the Admiralty's anonymous commentator wrote that:

> compared with peace practice, ranges of 10,000 metres (10,930 yards) and 12,000 metres (13,120 yards) sound preposterous, but they are not really so. Firing begins to look possible at 20,000 metres (21,870 yards), reasonable at 14,000 metres (15,310 yards), close range may be counted as setting in at about 10,000 metres (10,930 yards), and at 5000 metres (5468 yards) ships might as well be alongside each other as far as appearances and sensation of proximity go.[1]

The Royal Navy was struggling to extend normal gunnery range to 6000 yards.

Using the systems

Fire control was based on cooperation between the primary control party aloft and the range-keeper (clock operator) in the transmitting station.[2] Both had Dumaresqs. The party aloft, which could see the target, was headed by the control officer. He guessed an enemy course (hence a range rate), and either guessed a range or approved an initial rangefinder range. On this basis the clock in the transmitting station was started. The clock operator's Dumaresq showed him the tactical situation. Before firing began, he corrected the rate as necessary to bring the clock range into agreement with reported ranges. He kept the rate officer aloft aware of how reliable he found his ranges, ie, of how wide the scatter of reported ranges was, or of how consistent it seemed to be. That and his Dumaresq helped him evaluate rates applied to the clock.

It was never likely that the initial data were correct. Gunnery was a cycle of firing, observation, correction, and then firing again, much depending on how the observation was interpreted (spotting). The corrections themselves came to be called spots. The early British long-range firing experiments showed that hits could rarely be seen from the firing ship. Gunners had to depend on what was visible – the splash from a miss. It also turned out that misses beyond the target were often invisible, so the British learned to depend on what they could see. The British view (not shared by the US Navy) was that it was impractical to measure how far short a shell had fallen, so that corrections had to be by rule. It took time for a splash to form, so that a shell aimed properly for line (direction) might produce a splash near the target's stern.[3] A second lesson was that concentrated firing in salvoes was far easier to spot than individual shots, hence that central control was inescapable; by 1903 it seemed that spotting by individual gunners would be limited to 2000 yards (this figure later rose). Salvo firing with control from aloft seemed inevitable for ranges beyond about 5000 yards.

There was an essential caveat. The entire gunnery cycle was designed to hit a target following a steady course at a steady speed, not imposing high range rates, because otherwise it was difficult or impossible to relate a miss on one salvo to the target position on a later one. That made sense for two reasons. One was that the enemy would probably also be using something like a Dumaresq-Clock-spotting combination, hence could not hope to keep hitting if he manoeuvred violently. It seems to have been accepted by all navies before World War I that they had to keep hitting to achieve much. In the sole case of modern naval combat, the Russo-Japanese War, Russian ships succumbed to massive cumulative damage, not to a few devastating hits. The spectacular effects of single hits were always the deaths of particular key officers, and they could be attributed to poor Russian operating practices. That World War I proved otherwise was a surprise both to the British and the Germans.

Conversely, a ship could evade damage by manoeuvring, because each manoeuvre would throw off the opponent's fire-control solution. Off Samar in October 1944, for example, US destroyers frustrated Japanese battleship and cruiser fire by manoeuvring ('chasing splashes') so that corrections on each cycle drove the shells further from their targets.

The control officer was responsible for spotting corrections. By 1905 many ships had adopted a technique of bracketing (it became standard about 1909). If the first salvo fell short, a second was fired with a set up-correction (by 1911, typically 400 yards); if it fell over, the correction was brought down the set amount. If the target was crossed (the second salvo fell on the other side), the range was somewhere within the set correction, and the control officer closed in by correcting the other way with half the original correction. Given successive halvings, soon the target should be straddled (shells of one salvo falling on either side). If the salvo was dense enough, a straddle should include hits.

The 1913 Fleet Orders suggested that ships of the *King George V* and later classes, with range-keepers (Argo Clocks or Dreyer Tables) and numerous rangefinders could use brackets as narrow as 200 yards at 10,000 yards if conditions were good and all rangefinders agreed on the range.

Standard British practice was for each salvo from a dreadnought to be half the available guns (the other half, already loaded, could fire the next salvo). In 1909 an experienced British gunnery officer suggested that the ideal salvo should spread out enough (say 150 yards) so that many of the shells in it would hit. 'It would mean a tremendous gamble, either to hit with all guns or none at all, and most people prefer to shoot driven game with a shotgun and not a rook rifle partly, at least, for similar reasons.'[4] By 1911 the 150 yards had been accepted (three times the zone in which half the shells would fall, at 10,000 yards). The danger space of a 13.5in gun at 10,000 yards was forty-one yards, so there was a good chance that some of the five shots of a salvo would hit. During World War I, the British and the Germans each thought that the other would have made many more hits had its salvoes been looser.

In an example used at the 1911 Long Course on gunnery at HMS *Excellent*, the gunnery school, the first salvo was over, so the gunner ordered 400 yards down for the second salvo of the initial bracket. The second salvo was short: the target had been bracketed. Now the gunner halved his correction, so the third salvo was 200 yards up. If it was still short, it placed the target in a bracket 200 yards wide. The next correction, assuming the target was within a 200-yard bracket, was another 150 yards up. That placed the target within the 150-yard width of a single salvo, which should mean straddling. Had the gunner seen splashes from both salvos, he would have corrected up (increased range). If he saw none, both salvoes would have gone over, and he would have corrected down. The 1913 Fleet Orders warned against concluding too much from a single round seen falling short, because it might be wild. Once the range had been found, ships should fire as rapidly as possible to establish fire superiority and overwhelm the enemy.

The control officer could adjust the enemy bar on the Dumaresq on the basis of what he saw of target course and speed. The clock operator could reset both the range itself and the rate; typically range rather than rate was reset. For example, if at some moment the clock showed 9500 yards, it might be reset to 9700 ('up 200') without changing the rate (the clock would keep running, always showing 200 yards more than it would have before the change). The control officer could cancel range changes by an opposite order (eg, clock range

Sir Percy Scott invented director fire largely in response to the advent of HMS *Dreadnought*. Given her massive battery, the existing practice of assigning a control officer to each main turret was no longer practicable. Without a director and a master firing key, guns were typically fired independently, making spotting inaccurate. On the other hand, non-director fire was faster, as the director typically had to wait for enough guns to be ready before firing. Although the rationale for HMS *Dreadnought* was her firepower, which included her ability to control that fire at long range, her rig has often been criticised because it placed her primary control position on the foremast, squarely in the stream of smoke from her forefunnel. The rig was worked out by Captains Jellicoe, Bacon, and Jackson – ie, by DNO, his successor, and the Royal Navy's radio expert. The theory seems to have been that smoke would affect either the tall foremast or the short mainmast position, but not both. However, another factor would have been obvious to the gunners. It was essential to estimate the course of a target. Because it took time to set up a fire-control solution, such estimation would best be done by observing the ship's topmasts before she popped over the horizon. A ship with a single topmast would enjoy a considerable advantage, as her course could not be estimated at all until her hull was visible. Given her speed advantage, and her considerable firepower over arcs other than her broadside, *Dreadnought* could exploit exactly this possibility. Given the need to handle boats using a boom stepped from a vertical mast, any single-masted solution had to entail stepping the legs of the tripod around the funnel, as in this ship. Trials demonstrated the smoke problem, and a more conventional arrangement was adopted in the next (*Bellerophon*) class. Initially it was to have had a similar rig, but with the foremast legs reversed. Then a full-height mainmast was added. However, the attraction of the single-mast solution remained, and it was revived a few years later.

up 100, spot down 100). He was advised to accept clock range whenever he was unsure of where shots were falling. For his part, the clock officer could change the range by fifty yards without reporting (unless he had to make two such changes within a minute).

Misses (resulting in corrections) could be due either to range (ballistics) or to rate errors. Choosing the wrong one would cause the ship to lose the target altogether. When a control officer ordered '400 up,' it was up not from the original range, but from a corrected range taking the range rate into account. If the range rate was too far off, fire would be pulled off the target even after straddling. Shells would fall short if the set rate was too low, over if the set rate was too high. Thus the reaction to shorts after a straddle would be to open the rate. The error would be attributed to rate if the spotter had to make two corrections in the same direction to cross again or to straddle. In that case the next spotting correction should be range plus rate in the same direction. It would be added to the rate on the clock. Thus fifty yards/minute up would add to the existing setting of 200 yards/minute down to give 150 yards/minute down. Typical corrections were 100 yards/minute brackets for beam bearings, 200 yards/minute for ahead bearings. Corrections to the enemy course and speed were deduced from the rate change via the Dumaresq. The rule of thumb was to correct course for an enemy running on a roughly parallel course, and speed for an enemy approaching from ahead. Most cases, unfortunately, would be intermediate.

It was therefore important that the clock officer evaluate any proposed rate corrections. There was what amounted to a dialogue between rate-keeper and control officer, eg, 'a very good rate is "opening 230"' or 'a very unreliable rate is "closing 200"'. After 'tuning' the clock to a new rate ordered by the control officer, the clock officer compared the resulting clock ranges with the incoming stream of rangefinder ranges. When the two agreed, he reported to the rate officer aloft that the rate was correct, ie, that spotting should be for range rather than rate (or should react to a change in target course). According to 1913 Home Fleet instructions, many spotters were too prone to change rates rather than ranges – 'spotting for rate'. Ships with very small salvo spreads, such as the 13.5in ships, were prone to lose the target in this way. The clock officer could also propose a new rate based on a range-time plot. Plotting was a valued way of rejecting bad range readings, which otherwise could throw the system off by making it seem that the clock had been mis-tuned.

In 1913 the favoured emerging method of rangefinding was called rangefinder control, using a Dreyer Table.[5] It was essentially clock-tuning, except that it could employ more rangefinders (its plot made it possible to find a mean rangefinder range) and it was more highly automated. The feedback pencil line made it easy to detect and correct errors in range rate; very large deviations showed that the enemy was changing course. Limited enemy manoeuvres could be compensated for temporarily by imposing a set spotting correction, provided range rates were small. The fleet instructions carried the warning that the plot on the

table should not be used to set a rate until the plot had been certified as good and the control officer had agreed to turn to 'rangefinder control'. 'The degree to which the range plot can be trusted will depend on the spread of the rangefinders as well as on the experience of the officer in charge of the table.' In intermittent visibility the ship might have to use rate control, with minimum spotting. Overall, the instructions cautioned against relying too heavily on a table fed by data that might well be unreliable.

Few ships yet had Dreyer Tables. For them the best method was direct observation by rangefinder. However, visibility would probably be intermittent, and readings could not be continuous. An alternative was to apply a rate from a Dumaresq to a clock ('rate control'). A third method combined the first two. A rangefinder or spotting checked the accuracy of the range projections. Hopefully, once tuned, the clock would keep the range accurately enough for a few minutes of firing. Alternatively, a ship could fire so rapidly that spotting alone would suffice to hold the range. Suited mainly to medium-calibre guns, this method was described as a last resort for large ships. The 1913 instructions included Dreyer's cross-cut technique. Note that prior to the advent of the Dreyer Table and the Argo Clock, there was no direct connection between the clock and the range transmitters in the transmitting station. Instead, the clock operator called out range changes to the transmitter operators in the standard twenty-five-yard steps. That step imposed dead time and was a source of error.

The rangefinder itself could be used as a check on the range rate. By 1913 the Admiralty had asked that the Pollen rangefinder mount be modified with a small electric motor, which could be set at the assumed range rate. If the two images seen through the rangefinder remained in coincidence, the rate was correct. Similarly, the Admiralty asked that the rangefinder be subject to electric pointing, providing feedback in bearing rate. Like the range, bearing was transmitted to the plotting table by step-by-step transmission, in this case with a precision of a quarter-degree.

The director

Salvo fire required one officer to fire the ship's whole battery. Percy Scott, who had invented continuous aim (see chapter 1), recognised that the solution to effective salvo fire was more centralisation: the better the coordination between guns, the tighter the salvo, hence the more accurate the system as a whole. To this end he revived the old idea of the director, originally a means of concentrating and commanding a ship's firepower. The first British approach to the problem may have been an 1829 proposal by Carpenter William Kennish RN. It was tried aboard HMS *Hussar* at Bermuda.[6] If the guns were all trained to concentrate their fire at a chosen range, and set for the right elevation at that range, and all fired together when the ship was horizontal (at the top or bottom of the roll), she would produce the desired tightly bunched group of shots. To this end the moment of firing was chosen by an observer at a 'marine theodolite', whose telescope was kept horizontal by a pendulum. Because each gunner pulled his own lanyard, each imposed his own time lag between the call for fire and the moment of firing.

About 1868, however, the Royal Navy adopted electric firing. Now a single gunner with a switch could fire all the guns simultaneously. Kennish's theodolite was renamed the director. Guns were more powerful, but ships usually still had them in broadside positions to which the Kennish concept was well adapted. The guns' racers (turning circles) were marked with particular angles of train, to converge their shots dead abeam at a chosen range. Without convergence, guns at different points on a ship would not hit the same point at the desired range. The appropriate train angles depended on the range and also on the bearing of the target.

Elevation staffs were marked for set ranges, typically 800 and 1100 yards. By 1882 a setting for 1600 yards had been added, with convergence marks for bow, beam, or quarter targets. Ships had directors on either side of their conning towers.[7] In effect they were the ship's gunsights. Each contained a telescope that could elevate and train in a frame marked with the three convergence bearings, and with a pointer showing deflection, marked in target speed. The guns were fired

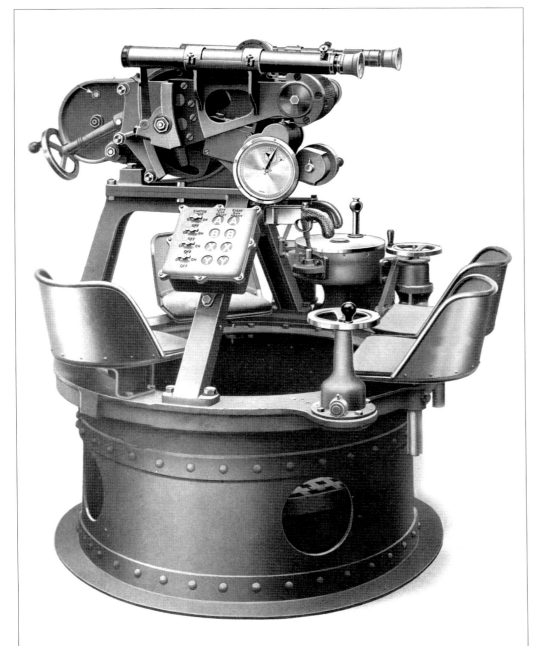

Scott's director is shown in developed form, on the tripod mounting used for battleships and cruisers. Destroyers had a similar director on a pedestal, without seats for the operators. This photograph is from the 1917 Director Handbook, at least one copy of which was given to the US Navy when the United States entered World War I. This director acted as a master or 'dummy' gun, the guns it controlled following its motion. Sights are set and the target is followed by pointing and training, as with a real gun. Because the sight on the director is set to the appropriate range, the guns automatically assume the appropriate elevation, including the inclination of the reference plane (through the ship's deck). The deflection of the master gun includes both deflection set on its sight and the relative bearing of the target as seen from the director. Individual guns are corrected for their distance from the master gun (parallax and dip).

when the target crossed the cross-hairs (horizontally and vertically). Waiting until the target passed through the horizontal cross-hair ensured that the ship was on an even keel when she fired. The vertical cross-hair provided the appropriate deflection. Although independent fire might be better at short ranges, arguably the director was essential when drifting smoke made it impossible for individual gunlayers to work effectively. This system was adopted in other navies as well, particularly for broadside ships.

The director defined a lethal patch on the ocean into which the ship could pour fire. Captain and navigator had to manoeuvre their ship so that this patch covered the enemy long enough to destroy him. As the official British gunnery manual of the time (1885) explained, the limited range flexibility was acceptable, because trajectories were so flat. Hitting depended mainly on whether the guns were fired at the appropriate point in the ship's roll. An accompanying diagram showed that guns converging at 800 yards would hit effectively anywhere out to 900 yards. Curves of danger spaces showed that for breech-loading guns rangefinding could not matter much below 1000 yards.[8]

The few preset firing choices inherent in the director system became less and less attractive as ship speeds increased (so the target might not remain in the lethal patch for a sufficient length of time). By 1893 the director was being eliminated from turret and barbette ships, ie, from ships with small numbers of heavy guns. Other ships retained leads from the guns to the conning tower, the director being, in effect, the captain's firing key.[9]

Scott seems to have realised that the fleet needed centralised control from a position remote from the guns, both to overcome smoke interference and confusion and to produce the necessary tight salvoes. Salvo fire by heavy guns was the only way to give them something like the precision which continuous aim offered lighter guns. As long as the guns could not follow the ship's roll, they had to be fired at the appropriate moment. Better to have that moment chosen by a single gunner instead of many, hopefully in a way consistent from salvo to salvo.

Scott later claimed that he had proposed a revived director as early as 1902, but that paper is now lost. He then concentrated on the need to remove control from the gun positions, so in 1903 he proposed a remote sight for one twin turret of HMS *Hero*.[10] Wires mechanically transmitted the motion of the sight to the guns. The *Hero* sight was rejected after trial, but it was later modified to become the first prototype director. In working out details, Scott became familiar with solutions to the problem of converging a gun on a distant target using an offset sight.

In 1905, having been appointed Inspector of Target Practice (in effect director of gunnery research) Rear Admiral Percy Scott proposed central control as the only way to handle the guns of the new all-big-gun battleship HMS *Dreadnought*. The system then developing provided each turret with its own spotter aloft. That was possible in a ship with only two turrets, but *Dreadnought* had four bearing on each beam. The single control officer aloft would have to treat the whole 12in battery as a unit. He would do better if he also had a master sight and a firing key, near the rangefinder, free of smoke and other interference.

This was the old director cured of its limitations. In the past, a limited number of combinations of range and target bearing had been preset. Scott sketched a device that would perform the necessary calculation at each gun turret, based on the range and elevation it received. Another device at the guns could correct for dip (the difference in height between director and guns). Later it turned out that turrets also needed corrections for the extent to which their roller paths were not precisely level with the ship. Other errors were associated with the fact that the ship was flexible; it bent and twisted slightly in a seaway. None of these complications was crippling.

Instead of defining a patch on the ocean, the director would act as a flexible gun-sight. Its telescope would be set for the appropriate bearing and elevation. When the sight-setter reported that the guns were ready, the director officer could fire the whole salvo together by firing key as soon as his cross-hairs were on the target, ie, as soon as the ship was at one end of its roll. This was what individual gunners already did. Scott understood that centralised control would reduce the rate of fire, perhaps by as much as a third; but, independent firing would produce ragged results difficult to spot, and the wind might blow the smoke towards the firing ship, making independent gunlaying impossible.[11]

The director so obviously fitted the spirit of the new centralised fire control that DNO ordered Scott's remote sight for HMS *Hero* modified as a director for the two right-hand guns in HMS *Vengeance*. The commanding officer of HMS *Excellent*, the gunnery training establishment, commented that the training gear of the ship (by this time, HMS *Colossus*) would probably not be quick enough for the director, although more modern turrets could be controlled successfully. DNC suggested using HMS *Dreadnought*, but Controller (Third Sea Lord, responsible for naval materiel) ruled that the ship was far too important for such tests. Ultimately (August 1906) the new pre-dreadnought HMS *Africa* was chosen. Referring to further delays by Vickers, a frustrated DNO (Captain John Jellicoe) wrote that 'it is very desirable to try this arrangement as soon as possible…this matter is very urgent….' A 4 March 1907 preliminary report on the *Africa* trials (which apparently no longer exists) was encouraging. An extemporised director for the 6in battery of HMS *King Edward VII* was described as quite successful.[12]

Scott's critics argued that raggedness in firing might be the least important factor in the spread of a salvo. Even when all guns were fired by a single key, they would experience different firing delays due, for example, to the variable rate at which the primer ignited, the powder itself, the details of ramming (ie, how well the powder and shell were pushed together), the wear of the gun, and the gun itself. While the gun fired, the ship rolled. The longer the interval between pressing the firing key and the emergence of the shell, the greater the effect of the ship's roll and other motions. A calculation for HMS *Dreadnought*, rolling four degrees each way within a period of seven seconds, firing at a target abeam at 7000 yards, showed a total error of 319 yards, well beyond the normal salvo spread.[13] Scott's rejoinder was that experiments on

Most battleships were fitted with directors under their foretops, as shown in this photograph of the modernised Chilean battleship *Almirante Latorre*. She had served in the Grand Fleet as HMS *Canada*, and was the last survivor of those battleships. The director is the cylinder below the foretop. The unusual circular house atop the fire-control top was a World War I fitting protecting a rangefinder. Visible atop the conning tower is the usual armoured hood for a 15ft rangefinder, with a second director inside. Modernisation between 1929 and 1931 entailed few visible changes, although it was always described as a fire-control refit.

board HMS *Good Hope* showed that near the middle of the roll the ship's motion was nearly uniform for long enough that all guns would suffer the same shift in range, which could be handled as a spotting correction. It was, moreover, much easier to recognise the middle of the roll than the top or bottom.[14]

While work on the prototype director proceeded, HMS *Jupiter* tested whether a telescope aloft could be held on a target, against the vibration of the mast. At up to 10,000 yards a seven-power telescope was held on the target, which crossed the cross-hairs (for firing) ten times a minute when the ship showed just the sort of irregular corkscrew motion (including significant yaw) that was expected to make it difficult to follow a target (13 January 1906). Rougher weather reduced firing opportunities (both vertical and horizontal cross-wires on the target) to about once a minute at 7000 to 9000 yards. *Jupiter* also trained her turret guns at fixed elevation to follow the target as she corkscrewed. The firing rate would have been slightly better than once a minute, a reasonable measure of how often a director ship could fire salvoes.

In his July 1907 report to his successor DNO (Captain John Jellicoe) pressed for director develop-

Organisation and Combat Information Centre (CIC). During World War I the British learned how difficult action plotting was. They had imagined that ships out of sight of each other could simply plot the positions determined by their navigators. The British discovered to their horror that the usual errors in navigation made such plots almost meaningless. At Jutland, Admiral Jellicoe remarked that his master plot seemed to show a cruiser squadron proceeding at three knots and Beatty's battlecruisers at sixty. He may have realised how lucky he was not to have adopted divisional tactics. By the end of the war, however, the idea of action plotting centred on the fleet flagship rather than on fixed positions had been accepted by the Royal Navy.

After Jutland British officers bitterly criticised Jellicoe for avoiding decentralisation and divisional tactics.[12] They thought the Germans had practiced both successfully (it is not clear to what extent German central command simply collapsed). They made no reference to the situational-awareness issue. The alternative to manoeuvering ships in divisions was to manoeuvre their fire (using concentration techniques). The longer the gunnery range, the more scope for such tactics. Jellicoe's successor Admiral Beatty chose to emphasise them.

Rethinking gunnery

The earliest surviving British war orders seem to be those prepared by Admiral Jellicoe when he commanded the 2nd Division of the Home Fleet in 1910–12.[13] Jellicoe planned to open at 15,000 yards (if weather permitted) and develop maximum fire at 12,000 to 13,000 yards. He expressly cautioned against going inside 7000 yards, for fear of torpedoes. Unfortunately Jellicoe's ideas did not match what the British thought their guns could do. Official war-game rules issued in July 1913 took 12,000 yards as the maximum range for 12 and 13.5in guns.[14] According to these rules it would take about twenty minutes for one *King George V* to neutralise another at 7000 yards, and about twenty-six minutes thirty seconds at 10,000 yards.[15] Matters would be considerably worse at Jellicoe's preferred battle range. The rules matter because they were essentially the ones used to evaluate success in frequent tactical (PZ) exercises. Thus they reflected what British naval officers thought would happen in battle. The rules imply that even at 7000 yards it would take too long to destroy an enemy battleship. Apart from his 13.5in superdreadnoughts, Jellicoe's fleet could not achieve much very rapidly at 10,000 yards. The whole fleet would have to fight well inside torpedo range.

In October 1913, Admiral Sir George A Callaghan, at that time Home Fleet commander, issued war orders envisaging opening fire at 15,000 yards (if weather permitted), closing to a 'decisive range' of 8000 to 10,000 yards where superiority of fire might be established. Ships might press home attacks at shorter ranges.[16] Callaghan arranged experiments specifically to determine maximum effective range (in the one set of experiments he ran, he tried for about 14,000 yards). A supplementary memorandum added after the 1913 manoeuvres defined the roles of the ships other than battleships: because the primary role of the battleships was to destroy the enemy's battleships by gunfire, the role of the rest of the fleet was to hold off any other enemy attacks – eg, by battlecruisers or destroyers – which might break up the British line and interrupt its firing. Much of this language can be seen in the later Grand Fleet Battle Orders, eg, the injunction that the enemy's battlecruisers are the primary target of the British battlecruisers once battle is joined (before that they are primarily scouts capable of breaking through any screen covering the enemy's movements). Callaghan's battle instructions also emphasised the role of British destroyers attacking the enemy fleet (the destroyers' role was developed in more detail in a March 1914 memo).

Presumably Callaghan's reduced battle range reflected the experience of the past few years. Instead of steadily increasing battle-practice range, the Royal Navy reduced it in 1912, from 9000 to 8000 yards. No explanation was given, but shorter ranges would have been consistent with the belief that North Sea conditions would make longer ranges irrelevant. Admiral Jellicoe later wrote that pre-war battle practices had been fired at up to 9500 yards, the only exception being a 1912 shoot by HMS *Colossus* off Portland in 1912.[17] To some extent the reduction was caused by a 1911 fleet performance so disappointing that a special gunnery conference was called to discuss remedies.[18] Strikingly, the conference suggested no technical remedies, only better training. Partly as a result of reducing range, the eight best ships made better than 30 per cent hits in 1912, whereas in 1909 the fleet average had been about 20 per cent.[19] The range reduction meant that existing gunnery techniques had reached their limit. It would take considerable investment to maintain the hitting rate while greatly raising battle-practice range – which measured the fleet's capability. Given North Sea conditions, that was probably not worthwhile. The fleet would usually fight at or inside typical battle-practice ranges. That was not to deny that, under good visibility, it might open fire at greater ranges.

Thus, up to the outbreak of war, British operating orders show a steady reduction in expected battle range, not the sort of continued increase which might have been expected given the efforts expended from 1904 onwards. This reduction, moreover, was accepted, even though torpedo range was constantly increasing. Trials in 1908 showed that a lengthened 21in torpedo could reach 10,000 yards at thirty knots, and Jellicoe assumed this performance in his 1912 tactical trials. The 21in Mk II that equipped British battleships in 1914 had a range of 10,000 yards at twenty-eight knots. The German G7 was credited with 10,000m at twenty-seven knots.[20]

Rangefinder control

The system of fire control the British had developed prior to about 1912 did not solve the 'browning-shot' problem because it could not demolish a German battleship before the latter's torpedoes had struck home. If the British found some way of pouring on fire more quickly, they could do sufficient damage and then carry out evasive manoeuvres. The German torpedoes might well not be visible, at least at any distance, so evasion would have to be carried out whether or not the Germans fired. Visibility and the need to evade set

drastic limits on the time available for firing. Unfortunately the record of policy decisions is very incomplete, perhaps intentionally so.[21]

Various official comments show that in the spring of 1914 the Royal Navy was developing (and emphasising) a new technique it called rangefinder control. The attempt to measure the range rate was dropped. Instead, a single fixed correction was taken to convert rangefinder range into gun range. Since the difference depended on range rate and time of flight (ie, range), among other things, no single correction could be acceptable for any length of time. The single correction would last longest at medium range (where error margins, given by danger spaces, were large), particularly if range rate was slow. Reliance on a rangefinder

There was initially confusion as to whether ships could or should be able to fire much of their broadside close to the centreline. HMS *Colossus* was one of the last British dreadnoughts with substantial firepower nominally facing forward, in the form of wing turrets. She and her sister *Hercules* were near-sisters of HMS *Neptune*. Commenting on special gunnery trials by *Neptune* and the battlecruiser *Indefatigable*, apparently the first to involve wing guns fired nearly fore and aft, DNO commented in January 1911 that blast effects were much more dangerous than had been imagined, involving not only deck shock but also objects thrown about violently enough to cause considerable injury. *Colossus* is shown at Scapa Flow, probably late in the war. She has a director on her fighting top, and one of the vertical objects on the bridge is an anti-aircraft rangefinder (the other is probably a tactical rangefinder seen end-on). Also visible on the bridge is a mechanical semaphore. All of her turrets show rangefinders; her main (stabilised) rangefinder is inside her fire-control top. *Colossus* was also one of the first class of British battleships with a single fire-control mast, the same feature which bedevilled HMS *Lion* (the otherwise very similar *Neptune* had two masts).

was possible because plotters could reject wild readings and (in 1914) they could average multiple rangefinder readings by eye. The single measurement of the correction would be made by spotting a few shots or a salvo. Rangefinder control also considerably simplified concentration fire, since there was no need for several ships firing at the same target to distinguish their splashes.

The Dreyer Table, with its range plot, was key to rangefinder control, and the new technique may have been the most important reason to adopt it.[22] As defined here, the technique was quite different from the rangefinder control (later called rate control) defined in Home Fleet gunnery orders in 1913–14; it was a very late pre-war bloom, not available just a few years earlier. That lateness was crucial to what happened once war broke out.

The key to volume fire was to abandon further spotting, at least until many shots were seen to miss. As Admiral Jellicoe pointed out early in 1915 (see chapter 3, page 80), under many circumstances independent gunlayers could fire the fastest, which was what was needed. His early war instructions envisaged rapid independent fire, slowed periodically to make sure that shells were still hitting. If they were missing, ships would spot to find new corrections, then resume rapid fire. Apart from target designation, there would be little need for centralised control. The new turret hydraulics made something close to continuous aim possible, even for heavy guns.

Rangefinder control explains the tactics Winston Churchill attributed to the pre-war Royal Navy in Vol III of his 1927 history of World War I, *The World Conflict*. First the British would smash the German battlefleet, then they would evade whatever torpedoes the Germans launched.[23] Evasion meant following a standardised signal, a blue pennant that ordered each ship to turn away from the enemy line. Ships' fire-control solutions would inevitably be ruined by so radical a turn. A turn towards the Germans would have exposed the fleet to more effective enemy fire (since it would also have closed the range) to which ships could not reply.

Churchill paid little attention to technicalities, but his account probably reflects what he was told when serving as First Lord before the war. Smashing means neutralisation. Even at the 10,000-yard range cited, it would take a German torpedo only about ten minutes to reach the British line – and the British (as personified by the then First Sea Lord, in November 1912) were well aware that such a 'browning' shot would have about a 30 per cent chance of hitting. Given this fact, why were they so optimistic on the eve of war? They must have had some way of solving their tactical problem.

Churchill associated the tactics he described with a half-hour engagement. That could not apply to the phase during which torpedoes were running. However, the half hour certainly could apply to a battle opening at much longer range, including an approach phase lasting as long as twenty minutes. During that time the British would begin firing, to achieve the 'fire superiority' Jellicoe sought. However, they would not begin to do devastating damage until the Germans came much closer.

Rangefinder control seems to have been very secret in 1913–14. Although in his April 1914 minutes DNO called it a key development, it was not included in the confidential Home Fleet Orders, which included extensive notes on fire control. The secrecy accorded rangefinder control, and DNO's claim that it was a very important new capability, suggest that the Admiralty had medium-range tactics in mind, but did not want to publicise a retreat from its previous emphasis on longer ranges.

There was an excellent reason for such secrecy. It was generally understood that the Germans considered their gunnery effective at medium range, about 6000 yards. Quite aside from intelligence information (which seems to have been excellent), there were many physical reasons to think that the Germans planned to fight at short range. The most commonly mentioned is the heavy medium-calibre armament of their dreadnoughts, intended to join in the main engagement. The trademark German tactic of passing destroyers through their battle line was associated with medium range because the German destroyers would have been chewed up by British shellfire.[24] As if to confirm British guesses, in his 1915 tactical orders (which the British obtained) the German fleet commander announced his intention to fight at 6600 to 8800 yards (presumably a translation of 6000 to 8000 metres).[25]

As long as the Germans did not understand that the British were building a medium-range trap for them with their new fire-control technique, they would rush into position to be demolished. If they held back, the British would have to fall back on the earlier techniques of deliberate fire, which promised only indecisive results (the Germans would not have done so very well themselves, but that was not the point).

It may be asked why, if the Germans were interested only in medium ranges, they learned how to fire effectively at the much longer ranges of Dogger Bank and Jutland. The answer lies in their analysis of how to get into range. Until about 1911 they seem to have thought that by adopting high range rates as they approached the British, they could avoid almost all damage. Then they realised that range rate might not be sufficient protection. If they could fire on the way in, while the range was changing, they might make British fire control ineffective.[26] In the autumn of 1914, British naval intelligence published the secret German report of gunnery practice for 1912–13.[27] The Germans had recently begun practicing long-range firing under 'difficult conditions', at ranges of 11,000 yards and above. The longest range for any of the capital ships was 15,000 yards. For example, the recent battleship SMS *Kaiser* had fired at 14,000 yards down to 13,500 yards at a closing rate of forty-three yards per minute. Typical results for heavy ships were 9.2 per cent hits at 12,000 yards.

The Admiralty's public position was that it was continuing to improve long-range performance. Not everyone bought it. A Russian naval officer wrote in 1914 that his service concluded from the British long-range gunnery results published in October 1913 that either long-range gunnery could not be effective (both sides would have to maintain steady courses) or that British interest in it was a cover for some alternative secret fire-control technique.[28] The Germans were acting as though long-range gunnery was ineffective, for example, by not providing ships with raised spotting platforms and by equipping them with only one or two rangefinders. The Russian thought that the Germans imagined they could reach their desired range without suffering much damage by closing at high speed, hence imposing a high range rate on the British (the Germans certainly did emphasise radical changes of course and high range rates in their exercises). The Russian thought the Germans had come to this conclusion as the result of four years of experiments using the armoured cruiser *Blücher*.

The Russians thought that the Admiralty was well aware of German thinking and, crucially, that:

> it is thought that German military principles and doctrines have become more acceptable and carry greater authority with your Board, than used to be the case. Many things combine to show that the present Board of Admiralty are no longer relying on the hope that future battles will be fought at very long ranges, and that their policy of adopting more powerful guns has, as its primary motive, the capacity to destroy an enemy by a single blow. That the heavier shell may give an advantage at long ranges in favourable conditions, appears to be only a secondary consideration. We hear also, that the German example in preserving a mixed armament will be followed. It seems to us, therefore, that the Board's refusal to try your [Pollen] system in its completeness can more easily be explained by the adoption of the German doctrine of short-range naval battle, than by your suggestion that they suppose that they possess a better long-range system. We have no indication that leads us to think, nor does the character or results of Battle Practice indicate, that the British system is superior to any other.

No surviving British official publication confirms the Russian's speculation. However, the war-game rules are very stark. The ranges used are not much greater than the Germans'. Even then, using conventional methods, no British ship fighting at such ranges could neutralise her opponent in time to avoid being torpedoed, if the Germans fired torpedoes at the same time the British opened fire. Yet the rules do not contemplate fighting at much greater ranges in order to avoid that threat.

War

When war broke out, the British had a tactical theory and a developmental fire-control system to match it, because they had not chosen rangefinder control until 1913 or even the spring of 1914. There was no effective long-range-gunnery alternative because, except for a few experiments, no effort had been made to practice firing much beyond 8000 yards, let alone to buy materiel specifically to support that option. When he took over the Grand Fleet, Admiral Jellicoe was thus placed in a very difficult position. The range at which the fleet had been trained to fight – 8000 yards – would have placed it in torpedo water without the hitting rate to win before the torpedoes arrived. Effective rapid fire at medium ranges was a future rather than a current proposition. His choice was to announce to the fleet that it would fight at maximum range. That announcement is what we now find in the Grand Fleet Battle Orders. There is no particular reason to imagine that ships and materiel that had struggled to hit targets at 9000 yards could suddenly hit at ranges 50 per cent greater.

When Jellicoe took office in August 1914, his draft tactical orders envisaged deploying at 16,000 yards, but opening range was dropped from Callaghan's 15,000 yards to a span of 9,000 to 12,000 yards (which he

Tacticians had to take torpedoes as well as guns into account. The Imperial German navy's torpedo boats (which included its destroyers) practiced a tactic of bursting through the middle of a battle line during a gun engagement. For a time this seemed flamboyant rather than realistic, as such craft could not expect to survive an onslaught of heavy-calibre shells. By 1910, however, the Royal Navy was noting that the Germans planned to take their destroyers to sea with them, and to use them during a gunnery battle. This idea seemed consistent with German insistence on seeking relatively short battle range, as the shorter the distance between battle lines, the better the destroyers' chance of surviving long enough to hit. The British had assumed that destroyers would present a threat only before or after the gunnery phase of the action. Hence they could mount their anti-destroyer guns in the open, their crews unprotected. British destroyers were not the answer, because they too would be chewed up by capital-ship shellfire. The *Iron Duke* class (the name ship is shown) responded to this new reality. She (and the corresponding battlecruiser *Tiger*) had protected 6in secondary guns instead of the earlier 4in. The greater weight of the battery and its protection forced it down into a much less favourable position closer to the water. Gunlayers there had a much more difficult time seeing their targets, so director control became more important for such batteries. Newly completed, *Iron Duke* shows a main-battery director atop her fire-control top. In this photograph she still lacks the big armoured hood for a stabilized rangefinder atop her conning tower, and only her B, Q, and X turrets show rangefinders. She was Admiral Jellicoe's flagship at Jutland.

described as long range). Jellicoe's orders suggest that Callaghan's longer-range test firings were less than successful and, moreover, that he expected decisive range to be something like 6000 to 8000 yards (no decisive range was cited in the draft orders). A few weeks later he issued a radically different Addendum No 1 to Grand Fleet Battle Orders envisaging opening with 13.5in guns at 15,000 yards and with 12in guns at 13,000, and at even greater ranges should the enemy fire first. Ships would shift to rapid fire at about 10,000 yards.[29] The orders were extraordinary. Almost nothing had been done to practice such firing.

Very little had been done to solve the short-range problem, either; rangefinder control was still nascent, and few ships had the Dreyer Tables they needed to generate reliable ranges. Jellicoe's orders make perfect sense as a temporary solution which would preserve the fleet until it was ready to fight on the terms that were being developed.[30] This logic makes sense of a February 1922 lecture describing the development of Grand Fleet tactics, delivered by Captain H G Thursfield to the British Naval War College:

> The heavy secondary armaments of the German ships, and the fact that they were known to practise attacks on an enemy engaged with the battlefleet by torpedo craft passing through the battle line, led to the firm conviction that the Germans would endeavour to fight a close-range action. So firm was this conviction that our whole tactics were based upon it.

> Yet this conviction, besides being unjustified by later experiences, was wholly mischievous, for it led to two further propositions, the acceptance of which had a very cramping effect on progress. The first of these we have already mentioned – the conclusion that, once battle was joined, there was nothing to be considered except the gunnery duel. The second may be stated as follows: 'Since the

Germans want to fight a close action, to do so must be advantageous to them; therefore we should endeavour to avoid it: we must develop and practise the game of long bowls.'[31]

The first proposition echoes British intelligence assessments of the period. The second reads as a somewhat sarcastic rationalisation of the tactics the Grand Fleet practiced in 1915. By 1922 Jellicoe was being roundly criticised for refusing to come to terms with the Germans (ie, coming to shorter range). Thursfield was standing in for Captain Plunkett (later Admiral Lord Drax), who often said that it was pointless to stand off at long range merely to avoid damage. Thursfield may also have been summarising Jellicoe's explanation *to the fleet* of why he was demanding that they adopt long-range gunnery rather than the shorter-range practices familiar pre-war. It was the demand for hits at 12,000 or 15,000 yards that was radical, after all, given the pre-war record. Both Jellicoe and Callaghan had to accept that, if they did fight at such ranges – which were considered very long in 1914 – the hitting rate would be quite low, because little had been done pre-war to prepare the fleet for such action.

Rangefinder control was described in detail in the official 1915 *Gunnery Manual* as a standard technique.[32] After both Dogger Bank and Jutland the Germans derided British plotting, which to them explained both their hesitation in opening fire and inability to follow their zigzag manoeuvres. It seems likely that the British were trying to obtain consistent range plots associated with rangefinder control. Certainly rangefinder control would have been ill-adapted to following enemy manoeuvres. After what may have been the only full test of rangefinder control (20 December 1915, by the fleet flagship HMS *Iron Duke*), Jellicoe wrote to Admiral Beatty that 'he wanted to see by actual hits that we are not living in a fool's paradise by firing so much by plotted results at long ranges.'[33] That is, most of the Grand Fleet's long-range fire was under rangefinder control. The system worked: *Iron Duke* made twelve hits out of forty rounds fired at 7500 yards using three-quarter charges (described as equivalent to practice at 10,000 yards). Jellicoe's Grand Fleet gunnery instructions for April 1915 stated that below 10,000 yards it was more important to begin instantly and to fire quickly than to be sure of hitting. Ships should start short and spot up until shells hit. This was essentially rangefinder control.

Rangefinder control was peculiarly well adapted to wartime conditions, in which opportunities for practice firing were scarce (almost non-existent for the battlecruisers). All its elements (except measuring the range correction) could be practiced in port: passing ships could be plotted, gunlayers could practice staying on target, and gunners could practice rapidly loading their weapons. Actual firing was needed to practice alternative techniques involving spotting, since at some point spotters had to see real splashes. This distinction explains why Jellicoe was relying on a technique that had not yet been tested by firings. Rangefinder control helps explain the emphasis on rapid fire. Its range limitations help explain the poor gunnery by Beatty's battlecruisers at Jutland.

After Jutland the Germans commented that the British had waited too long (for their plots to form) before opening fire. Such wartime references suggest the use of rangefinder control: plotting was apparently intended to find a stable average rangefinder range for comparison to gun range. The comment (by a US officer) that at Jutland range rates were set to zero also suggests the use of rangefinder control, in which the Dreyer Tables would have been used only to average and evaluate rangefinder data. Despite repeated injunctions not to wait for plots to form, apparently gunnery officers preferred to wait: the 1915 Grand Fleet Gunnery and Torpedo Orders re-emphasised the need to open fire instantly. The Germans also commented afterwards that reliance on plots made it difficult for the British to deal with radical changes of course. By 1918 the Royal Navy agreed (see chapter 6).

Rangefinder control was a much more sophisticated reversion to pre-calculator ideas of control, with the understanding that no solution would be valid for long. It had the unintended consequence of encouraging officers to wait to fire until an acceptable plot had formed (ie, a reliable idea of the range had been

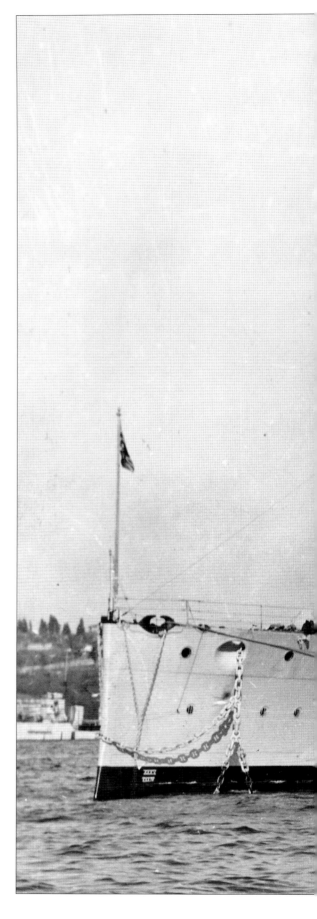

At Constantinople on 15 December 1922, *Iron Duke* shows wartime modifications such as a much-enlarged director top to her fighting top and a screened platform above her bridge for a tactical rangefinder. Tactical rangefinders, often described as torpedo rangefinders, were needed to help maintain a consistent plot of nearby ships' positions. In the Royal Navy the flagship's plot provided a fleet commander with vital situational awareness and thus with the basis for complex manoeuvres. Without a plot, he had to keep the evolving tactical situation in his head, an impossibility for a modern fleet spread out over a wide area, particularly when visibility was poor. Plotting was difficult, and not until late in World War I did the British realise that ships had to plot relative to the fleet flagship rather than use their own navigation. Fleets that did not use plotting were in a far worse position. It appears, for example, that the Imperial German navy never devised tactical plotting techniques, which suggests that at Jutland Admiral Scheer had only a limited idea of his situation. That helps explain why simply extricating his fleet from the danger in which it found itself was so high a priority. At this time *Iron Duke* had 25ft FX2 rangefinders in A and X turrets, 9ft FT8s in B and Q, an 18ft FT24 or 25 in Y turret, two 9ft FQ2 on her forebridge, another 9ft FQ2 on top of her chart house platform, a 15ft FT24 atop her gun control tower (with a 9ft FT8 inside), a 15ft FT24 on her torpedo control tower aft, and an anti-aircraft rangefinder (6.5ft FT29) on her after superstructure.

gained). Those who had invented rangefinder control wanted gunners to be able to open fire instantly and to pour it on once opened. This combination explains Jellicoe's exhortations to his fleet, in his Gunnery and Torpedo Orders (1 October 1915). He warned against waiting for the plot. Sometimes rangefinders would be useless, as at Heligoland. They were not the essence of the fire-control system. 'A good spotter with a thoroughly effective control organisation to help him should be capable of obtaining good results under any conditions if he is not over cautious.'

Particularly at long range, the gun might often be both rangefinder and range-keeper: spotting was more important than ever. In fine weather range might be limited only by the settings on the sights. Jellicoe did not intend to open fire beyond 18,000 yards unless the enemy did so, or the conditions required it – as in a chase, which occurred at the Dogger Bank. Ships must always be *ready* to fight at extreme ranges. Jellicoe's April 1915 gunnery orders show that to do that he relied on the pre-war technique of bracketing (ie, seeking to place salvoes on either side of the target, then correcting until the target was being straddled) to find the range, which meant not firing the next salvo until the first had splashed and been spotted.[34] Thus, at 18,000 yards a 13.5in ship could fire salvoes at fifty-second intervals (at 12,000 yards, forty-second intervals). Once the range had been found, ships would shift to 'rapid salvoes,' the next salvo being fired as soon as guns were ready. Any spotting corrections would be applied to a later salvo. Heavy guns could fire at least two such salvoes per minute. Jellicoe's April 1915 instructions envisaged accelerating to a salvo every twenty seconds once spotting could be discounted (as in rangefinder control). The maximum range envisaged was at least twice the pre-war range. Except for directors, nothing had been added to the mix to improve performance. Jellicoe later wrote about how hard he had worked the fleet to improve its long-range performance. He was fortunate in having the Moray Firth in which to fire (Beatty's battlecruisers had no such practice area). In 1915 they fired at least twice at 16,000 to 7,000 yards.[35]

The post-Jutland British 'gunnery revolution' (see chapter 5) was a long-range complement to rangefinder control, with the same emphasis on quick results and the same discounting of range rates and computation. The spotting and correction process envisaged pre-war was collapsed to find firing solutions which might briefly be valid, despite enemy manoeuvres. Anything more akin to pre-war attempts to find and integrate rates would probably be frustrated by those manoeuvres; no solution would last for long. Quick rate estimates based on Dumaresqs would have to suffice. That is why spotting rules, rather than computing techniques, were emphasised after Jutland.

The Grand Fleet Battle Orders assigned the two leading pairs of British battleships to concentrate on the two leading German ships (if the fleets were on opposite courses, the two rear pairs would attack the German van). This seems to have been a new feature, not taken from Callaghan's orders. The idea that concentrated fire could break up the German formation became more important after Jutland, and was an important postwar theme in British naval tactics.

For the British, the great wartime surprise was that the Germans never wanted to come to grips with them and risk a decisive battle. Jellicoe did not realise that this was because its masters periodically told the German fleet not to risk losses. He assumed instead that the Germans were trying to draw him into a mine and torpedo trap; he knew that they emphasised underwater weapons, and he felt that his own ships were deficient in underwater protection.[36] The 27 October 1914 loss of a modern battleship, HMS *Audacious*, to progressive flooding after being mined presumably reinforced this fear (however, the one British battleship torpedoed at Jutland, HMS *Marlborough*, survived). Jellicoe raised this possibility in an October 1914 letter to the Admiralty, and a 1 December 1914 Admiralty Confidential Order said that the Germans might invite pursuit specifically to draw the British fleet into such a trap, and that they had practiced this tactic. Jellicoe further emphasised this possibility in a December 1915 addition to his Grand Fleet Battle Orders; exercises both at sea and on the game board had shown that a German turn-away would be difficult to counter, because it would force him to choose between losing contact and losing effective firepower until he could reform his line.

Jellicoe's early wartime correspondence is bleak, probably because the new medium-range fire-control technique (ie, rangefinder control) was not yet fully available. He considered his ships inferior to the German ships on the whole, their only advantage their gun power. Without the new technique, Jellicoe almost certainly doubted that he could achieve decisive results. If his fleet remained within range long enough to accumulate enough hits, it would probably be vulnerable to long-range (albeit slow-running) German torpedoes. He could not do well enough at long range to gain decisive results at Dogger Bank. Through the winter of 1914–15 Jellicoe came to accept that the best the Grand Fleet could do was bottle up the Germans, as Togo had done to the Russians after the Yellow Sea. That had proven a worthwhile kind of victory. Jutland probably reinforced this view: as an American correspondent put it, 'the prisoner has assaulted his jailer, but he is still in jail.'

Apparently the Germans had no such plans for mass torpedo attacks. In effect they practiced that part of their tactics designed to get them into fighting range without being destroyed on the way, then stopped closing well before getting there. Thus battle ranges were almost always much longer than the Royal Navy had practiced pre-war. The pre-war medium-range concept became an embarrassment to the Royal Navy and Jellicoe who had championed it. Because rangefinder control went nowhere (at least for major units), it did not figure in accounts of fleet development. There was no incentive for anyone involved to write about the decision, in 1913–14, to emphasise medium-range fire. It is visible only in a few pages in the 1915 *Gunnery Manual*, and in conspicuous gaps in what would seem to be necessary preparations for the sorts of range capabilities demanded instantly once war had begun. The Germans had never expected to achieve much en route to decisive range. Thus the destruction they wrought at Jutland was a surprise as much to them as to the British.

CHAPTER 5
The Surprises of War 1914–18

The *Majestic*-class predreadnought HMS *Prince George*, used in the Dardanelles campaign, shows an early form of anti-rangefinder camouflage – the spirals around her masts – at Salonika in December 1915. They proved useless, and were soon removed. A small spotting platform was added at the very top of the foremast. She had had howitzers mounted atop her turrets in April 1915, but they were removed soon after the Gallipoli landings. Instead, in September 1915 A turret was fitted with a 12pdr field gun on an anti-aircraft mounting, barely visible here. The objects on the bow are for minesweeping.

World War I was a perfect example of the old saying that plans rarely survive contact with the enemy. It was a horrible surprise. Battlecruiser and cruiser battles were anything but line-ahead slugging contests. Ships manoeuvred individually and, often, radically. North Sea mists did not limit ranges as much as had been anticipated, but they often made visibility very intermittent even when maximum ranges were long. Targets often popped up quickly and then vanished into the mist, charging off at different courses that plotters could not follow. The Germans refused to let their relatively simple fire controls govern their tactics, so they made few hits. This behaviour could not easily be reproduced for training. No towed raft could manoeuvre. The British solution was 'throw-off' firing, in which guns were aimed to miss a real manoeuvring ship. The battleship-launched torpedo, so feared pre-war, proved irrelevant; few were fired at Jutland, and none hit. Destroyer torpedoes, which were easier to deploy against a manoeuvring ship, were a very different proposition. It probably surprised the British that their destroyers were far more successful than the German, even though they credited the Germans with far more interest in torpedo tactics.

In response to the Germans, in 1917–18 the Royal Navy revolutionised its gunnery. The British now accepted that firing opportunities would be brief and hitting rates low. The two main improvements in gunnery technique were a new kind of long-range control and effective concentration fire by up to four ships. The new control technique in effect applied the ideas of rangefinder control to long-range fire. To do that it needed a quick way of finding the range. Bracketing was much too slow. The Royal Navy adopted a new ladder system inspired (erroneously) by the Germans. Waiting for the shots to land in the water took time, perhaps up to half a minute at battle ranges. In a 'ladder', multiple salvoes were fired in rapid succession before any spotting was done. Those advocating ladder firing pointed out that despite its apparent wastefulness it saved ammunition by getting onto the target much more quickly.[1] Enemy zigzagging guaranteed that no solution would be valid for long.

Ranges were far longer than had been imagined pre-war. At the time of Jutland, the Grand Fleet Battle Orders called for opening fire (in good visibility) at 15,000 yards, and staying outside about 14,000 until enemy fire had been overcome, then closing to 10,000 yards to do decisive damage (10,000 yards was also the range associated with rapid fire). Orders issued soon after Jutland called for opening range to be within the capabilities of 12in guns but outside the danger zone defined by enemy torpedoes with 15,000-yard range; the van was not to close within 16,000 yards in high visibility.[2] The edition issued on 12 May 1917 called for all ships to be able to open fire at maximum range, which might be difficult to know under bad rangefinding conditions. Ships with 12in or smaller guns would open at 1000 yards beyond their maximum range as long as it was certain range was decreasing. These instructions emphasised the use of plotted ranges (ie, rangefinder control) after opening fire, as observation of the fall of shot was uncertain and could be misleading. If fire was opened before any rangefinder readings had been taken, this caution applied 'with greater force'. January 1918 orders re-emphasised the idea of concentrating on the enemy van in order to disorganise his fleet. There was a clear preference for long range, but the torpedo threat might be disregarded in order to make gunfire effective.

HMS *Queen Elizabeth*, flagship of the Grand Fleet, is shown in 1918, having incorporated all wartime fire-control improvements. This is a detail from a photograph taken from USS *New York*, flagship of the 6th Battle Squadron, the US battleships operating with the Grand Fleet. Note the anti-rangefinding baffles on the topmast. Apparently the baffles had been removed from the funnels by the time this photograph was taken, probably just after the war. The concentration dial (clock) below the fire-control top is not visible here. However, the associated bearing markings can be seen on B turret. This class was designed from the first to have its armoured stabilised rangefinder above the conning tower. The rangefinder hood also accommodated a director. This combination was repeated in the *R* class, in the battleship *Canada*, and in wartime *Renown*-class battlecruisers and the 'large cruisers' of the *Furious* and *Courageous* classes. A second (primary) director was installed atop the foretop, close to the gunnery officer and to the spotters. In the top, looking out through its windows, was a 9ft rangefinder (unstabilised). The identity of the stub above the aloft main-battery director is unclear. The main battery could also be controlled from B turret. This class introduced 15ft rangefinders, both in the armoured hood and atop the turrets. The official British rangefinder handbook (1921) credited this class with 30ft FX2 rangefinders on their high turrets and 15ft FT24s on their low turrets. In addition, they had a 12ft FQ2 in the foretop, 15ft FT24s in their armoured rangefinder towers (atop the conning tower) and on their torpedo-control towers aft, two 9ft FQ2s on the forebridge (for secondary-battery control), and a 6.5ft anti-aircraft rangefinder on the after superstructure. *Warspite* also had an 18ft FT24 or FT25 atop her spotting top. Directors for the secondary battery were first ordered in December 1914, but they had a lower priority than main-battery units for other ships; *Queen Elizabeth* received a temporary installation in November–December 1916, replaced by a full system the following March. The secondary directors were on each side of the compass platform. One is visible here, as a squat cylinder of roughly the same diameter as the main-battery director aloft. The short rangefinder visible above the compass platform was for tactical plotting. A torpedo director tower, with its own rangefinder on top, was located on the after superstructure just forward of Y turret.

Although the orders were not explicit on this point, by 1918 ranges of 20,000 yards – more than double the pre-war standard – were considered normal (ships typically exercised at 15,000 to 18,000 yards, and sometimes at up to 24,000 yards). By 1917–18 the Royal Navy was experimenting with air observation and spotting, and seeking ways to incorporate aircraft data in its fire-control systems.[3] Unlike surface observers, airborne spotters could see over the mist and they could often distinguish enemy courses directly. They could also see overs as well as shorts. This technique was limited, not least because airborne radios were not yet reliable. It was a pointer to the future.

The post-Jutland effort began with the first standardised British spotting rules.[4] The minimum ladder was a double salvo, shots being spaced just enough to avoid interference between the two guns of each turret. Ships normally opened with a double salvo fired either at the calculated range or, preferably, split 100 yards up and down (very few ships straddled on the opening salvo). Because it was vital to get line (bearing) right, standard practice was to split the double salvo in deflection, each half one-third to the side of the expected target bearing. If either or both of a double salvo straddled the target, the ship would immediately switch to rapid fire. If the entire double salvo fell short, then the control officer

HMS *Ramillies* shows wartime changes in this 1918 photograph from the P W Yeatman collection (US Naval Historical Center). Her B turret shows the bearing markings adopted in the Grand Fleet to promote concentration fire, in which one ship of a group might see a target not visible to others. However, she lacks the usual clock-like concentration dial. Her foremast shows a director, and a rangefinder was concealed in her foretop. Atop her conning tower is another rangefinder in an armoured hood. The housing atop her foretop probably protects a vertical anti-aircraft rangefinder. Unlike many contemporary British battleships, she still retained the very simple bridge structure with which she had been built. One important feature not visible here was a new version of the Dreyer Table, initially designated Mk V.

The older British dreadnoughts were dramatically modified. HMS *Bellerophon* is shown here at Scapa Flow in 1918 with bearing markings (for concentration fire) on her A and Y turrets, 4in anti-aircraft guns aft, and relocated searchlights. A concentration dial was added to the foretop in 1917–18, but it is not clearly visible here (it is probably the vertical object in front of the top, the dial being in profile). A director has been bracketed to her foremast, and a forward funnel cap reduces smoke interference with the maintop. The enlarged V-front navigating bridge was fitted about 1913, with the structure below it (which was enlarged in wartime). The two platforms above were wartime additions (*Temeraire* of this class had less extended bridgework abaft her conning tower). The 4in anti-destroyer guns atop the centreline turrets were relocated in 1913–14 to the superstructure to concentrate them. The guns on the beam turrets soon followed. Refits in 1914–15 concentrated the secondary guns in groups in the two superstructure masses. Searchlights were also concentrated, to make it more difficult for attackers to judge the ship's course from their light at night. In 1917–18 'coffee-box' control towers were built below them, so that operators would not be blinded by the lights. Director firing for the main battery was fitted in 1914–15. The 4in anti-aircraft gun aft was relocated from X turret to the quarterdeck in 1918. Some time later flying-off platforms (not visible in this photograph) were added to the end turrets. In effect this class was a production version of HMS *Dreadnought*, with the same layout. Note the kite balloon overhead, with its observation basket; *Bellerophon* was fitted with kite-balloon equipment in 1916 as a way of extending her horizon, partly for longer-range gunnery spotting.

would push a series of double salvoes up in range (spacing 400 yards). Similarly he would push them down in the event that the double salvo fell over the target. Either procedure was called a *ladder*. If each salvo had a 200-yard spread, a ladder of two double salvoes, each covering 400 yards, would cover 1400 yards in range (two sets of double salvoes spaced 400 yards apart).[5] Once the target had been crossed, a 200-yard correction would bring the double salvo back over the target.

The 1916 rules went into considerable detail about how to regain the range if the target manoeuvred (zigzagged) out of trouble. In effect, a radical manoeuvre suddenly changed the range rate. A ladder equated to a spread of range-rate choices, hence could compensate for such evasion. The 1916 rules referred particularly to the problems of light cruisers, but much the same might be said of a battlecruiser-on-battlecruiser battle such as the Dogger Bank of 1915.

Ladder ranging effectively extended the rangefinder-control concept. As in rangefinder control, the most important virtue of the Dreyer Table was not its ability to measure rates, but rather its ability to maintain a summary of available data, so that bad data could be rejected and ranges quickly averaged.

The sheer size of the table made modification relatively easy.[6] However, by 1918 it was criticised because it could project ahead neither own-ship nor target position. Nor was it well adapted to taking account of measured or estimated target inclination, an urgent matter if the enemy's course could change very suddenly.[7] When the US 6th Battle Squadron joined the Grand Fleet in 1917, the British came into contact with its synthetic fire-control computer, the Ford Rangekeeper. One was installed for tests on board the light cruiser HMS *Cardiff* in 1918, but no account of trials seems to have survived.[8]

Director control proved vital. At wartime ranges, and in poor wartime visibility, centralised fire control from an elevated position became essential rather than attractive, and the elevated position was probably also the only way to spot splashes at long ranges. By 1916 director installations for 6in secondary batteries were beginning to be put in place. After Jutland director control was extended to all battleship secondary batteries,

Off New York City in 1919 Grand Fleet veteran USS *Texas* displays three important wartime innovations: the aircraft atop her B turret (at this time, generally a fighter to deal with enemy reconnaissance by Zeppelin) and provision for concentration fire, in the form of bearing markings on A and B turrets and a clock-like concentration dial on the foremast. The bearing markings disappeared from US and British ships by about 1925, but the dials remained through the interwar period, and were adopted by other navies as well. The big rangefinder atop the bridge was also, at least to some extent, a fruit of wartime experience, as it was needed to support tactical plotting. In the case of *Texas*, it was a replacement for the long-base rangefinder previously mounted on B turret, displaced both by the airplane and by wartime heavy weather damage. US ships were also given permanent enclosed bridges, both for ship-handling in rough weather and to shelter plotting facilities.

then to cruisers, and ultimately to destroyers. The need for the director had not been at all evident pre-war.

Ships had to be able to keep hitting even if the target was obscured. That required the artificial reference that a gyro could provide. After Jutland the Royal Navy adopted Henderson gyro firing gear. Using a gyro (at the director) as a vertical reference, it fired guns when they were elevated for the proper range, ie, when the ship was at the right roll angle. A gyro alone would not have sufficed because it would have wandered. Instead, Henderson used a gyro-stabilised prism in the trainer's telescope. The trainer could see it wander as the crosshairs moved slowly up and down, and he could correct accordingly. The Royal Navy had tested and rejected the somewhat similar Austrian Petravic system (which the Germans ultimately adopted) in 1909.[9] Henderson gear became Gyro Director Training (GDT) gear.

Like all earlier British gunnery measures, the Henderson device stabilised *along the line of sight*. That there was no provision for cross-roll particularly afflicted the British battlecruisers when they chased the Germans. It was diagnosed only after the war; few if any of those involved pre-war had expected chases in which ships would roll heavily, hence throw their shots off for line. Apparently the Germans, unlike the British, had some limited provision to correct for cross-roll.[10]

Two-ship concentration had been tried pre-war. In bad visibility at Jutland, when up to four ships found themselves in sight of one or two enemy ships, they were unable to concentrate their fire. Concentration became more important as firing opportunities became briefer. On 27 February 1917 the Grand Fleet tried four-ship concentration for the first time.[11] If ships could distinguish their splashes, they could fire ladders simultaneously, quickly searching an area and getting onto the target. In the exercise, the ships fired in sequence, at fifteen-second intervals, using a range given by HMS *Colossus*, acting as master ship. All shots were short, so she ordered a 200-yard 'up-ladder', each ship firing in turn, the ladder being plotted. When more shots fell out of range, she ordered a 'reverse-ladder' spread 100 yards to either side of the range, corrected for ships' position in line and rate. Each ship plotted the fall of her own shots. The exercise showed that a 400-yard ladder would be quicker, but that the 200-yard ladder gave more hits earlier. If the ladder crossed the target, two of the salvoes would probably hit, whereas a 400-yard ladder could cross without any hits. Ships distinguished their splashes by firing only in assigned time intervals, using specially marked stopwatches. Director firing made this possible, since salvo duration (and hence splash duration) could be limited. Special radios (Type 31) were installed specifically to communicate concentration data. Ships were fitted

The battlecruiser HMS *Inflexible* shows typical wartime and late prewar modifications. Above her spotting top is her stabilised rangefinder, in its own cylindrical housing. It is the 15ft type introduced after Jutland. The spotting top itself carries a concentration dial, barely visible on its forward side, to show a ship forward of her the range at which she is firing (concentration dials are more visible on the light-cruiser mast in the background). The accompanying visual means of transmitting bearings was the set of bearing markings visible on the side of A turret. The foretop was enlarged, as shown here, in 1917–18. Below the spotting top is a platform carrying the ship's 12in director, in a cylindrical canopy. The bridge has been considerably enlarged. The 4in guns originally carried on the turret tops were moved into new positions in the superstructure during a refit in Malta in March 1915. The rangefinder atop the compass platform was added in 1918 to provide ranges for tactical plotting, a vital means of maintaining what would now be called situational awareness. The need for a protected plot helped force up the size of ships' bridges late in World War I.

with concentration dials showing the ranges at which they were firing, and their turrets painted to show their bearings.

The British had assumed that any night battle would be a mêlée, favouring the weaker side (the stronger side would suffer more friendly-fire errors, for example). On this basis Admiral Jellicoe declared that he would avoid night action. To some in the Grand Fleet, German successes the night after the battle demonstrated such superiority in night fighting that the Germans might actually welcome a night battle in future. Addressing this deficiency became a major postwar theme.

Heligoland

The first battle of the war, off Heligoland, 28 August 1914, was fought in bad visibility, so ranges were short: the first straddle came at about 5000 yards.[12] The British concluded that the ships had been badly hampered by their lack of director control: gunlayers using narrow-field periscope sights (adopted to protect them from blast) found it difficult to find and track their targets. Spreads were therefore larger than expected. A decision to abandon installations in new ships was reversed (a telegram in the Heligoland gunnery file directs the shipyard to suspend removal of the director from HMS *Benbow*). Other telegrams in the file say that 'director firing is [now] of the greatest importance.'

The Falklands

In December 1914 two British battlecruisers fought Admiral Graf von Spee's Pacific Squadron, which had recently sunk the HMS *Good Hope* at Coronel.[13] This time visibility was excellent (only in the last hour of the battle did it fall to 15,000 yards), and both squadrons steamed at high speed. Neither British ship had a functioning director or a Dreyer Table. Both ships found their fire control hampered by funnel smoke, so that although her fore conning tower and A turret never lost sight of the enemy, in *Invincible* the fore top occasionally lost sight, and P, Q, and X turrets were much affected. Rangefinding was very difficult due to the long range, funnel smoke, splashes and spray from the enemy. Rate-keeping was difficult at best, due to the enemy's zigzagging as well as to the very long range (variations in range were almost undetectable). Gunners found it difficult to stay on a point of aim, sometimes mistaking the target's bow for her stern. On the other hand, according to prisoners, British shells performed well, penetrating and exploding deep in the ships. Even so, *Gneisenau* took fifty 12in hits before sinking. It was no great surprise that the British ships used up most of their ammunition: one 12in gun in *Inflexible* fired 109 rounds (the ship was designed to carry eighty for that gun).

To the surprise of the British, von Spee's ships zigzagged to avoid being hit, even thought that made thought, wrongly, that they were using directors), and their direction and fire discipline were excellent. The British were impressed by the effect of plunging shells at such ranges, and by the blast effect of the German fire. The German survivors stressed, and the British noted, that slow British fire made it easier for their own gunlayers. It also made spotting easier, because the British ships were much less completely enveloped in the smoke of their own guns. This may have been the first of many British observations that their firing techniques were far too deliberate.

The Scarborough Raid

At about the same time as the Falklands engagement, the German battlecruiser force raided the British seaside towns of Scarborough and Hartlepool (16 December 1914).[14] British code-breaking had made it possible to send out an interception force, built around the 2nd Battle Squadron, the previous day. The battle squadron came close enough for some ships, but not the flagship, to sight the enemy in misty weather, which turned out to be typical of the North Sea. Apparently the pre-war Royal Navy had never appreciated that North Sea conditions did not limit range so much as they limited the duration of firing opportunities. Much depended on individual initiative – and on very quick reactions. Neither was shown off Scarborough. No one in the battle squadron saw fit either to open fire or to inform the squadron commander (they probably assumed that he could see what they saw). Some of the cruisers that did engage German light forces broke off when they received a recall signal, never asking whether the sender realised they were in action. None of them sent position reports.

Later, the official historian Julian Corbett described Scarborough as the worst naval shock of the war, because it demonstrated such fundamental weaknesses. It may have helped demonstrate to Jellicoe that the fleet was entirely unprepared for divisional tactics. Officers who did send contact reports generally did not include their positions. Without frequent position reports, plotting and hence situational awareness were impossible. Jellicoe felt compelled to issue a formal

Laid up after World War I, the battlecruiser HMAS *Australia* shows typical modifications: the stabilised 15ft rangefinder atop the spotting top (which also contains a 9ft rangefinder) and the director under the spotting top, on a foremast bracket. Her navigational rangefinder (which is end-on) is barely visible on her compass platform. The structure emerging from the back of A turret is a short rangefinder. This photograph was probably taken in 1923 (she was laid up in December 1921).

hits by their own guns unlikely. As crack gunnery ships, the Germans were expected to fire at maximum range, but the actual figure for their 8.2in guns, 16,000 yards, seems to have surprised the British. The Germans straddled (without hitting) at 15,000 yards. The Germans persistently fired salvoes (the British

The battlecruiser *Princess Royal* is shown after World War I, probably in the Reserve Fleet, with her turret bearing markings painted out, but retaining the anti-rangefinding baffles (added 1916–17) on her fore topmast. The bump barely visible at the after end of B turret is its 9ft rangefinder, whose ends did not project beyond the turret sides. An armoured rangefinder (in an Argo tower) is visible atop her conning tower. The British 1921 rangefinder manual credited this class with 9ft FT8 rangefinders in A, B, and X turrets – and a 25ft FX2 in Q turret (a drawing of *Princess Royal* showed a 30ft rangefinder there). The pre-war and early-war programmes had left the ships with 9ft FQ2 rangefinders in their foretops and in their stabilised Argo mountings. To these were added 15ft FT24s on the Argo mounting and on the torpedo-control tower aft, plus a 6.5ft vertical anti-aircraft rangefinder atop the foretop. The director in the platform under the control top was added in 1915 after Dogger Bank. It required a stiff support, so the reinforcing tripod legs were added at the same time. The concentration dial on the fore side of the fire-control top is not visible, but was surely in place at this time. Note the splinter-protection mattress on the compass platform, and the small rangefinder (for tactical plotting and torpedo control) atop it (added 1918–19). *Lion* had a high-angle rangefinder (with a vertical barrel) atop the fire-control top. The control top was enlarged in 1917–18 and the concentration dial fitted at this time. The additional control position atop the main fire-control top carried the anti-aircraft rangefinder, which appeared as a vertical stub mast. The small rangefinder visible on the compass platform was a navigational or tactical unit associated with tactical plotting as well as functions such as secondary-battery control and torpedo control.

order enjoining officers to show proper initiative when in touch with the enemy, and assigning tactical command in that situation. Similar problems recurred, a famous instance being a British cruiser that came upon a German cruiser the morning after Jutland and failed to open fire. As for the reports themselves, at Jutland, Jellicoe would complain that his plot was pointless because, if he took it literally, some ships were moving at sixty knots and other at three.[15]

Dogger Bank

The next engagement, this time between battlecruiser forces, was Dogger Bank (24 January 1915).[16] As in the Falklands, visibility was excellent, so the battle was fought at ranges not imagined pre-war. It was a wake-up call for both fleets. Vice Admiral Sir David Beatty, commanding the battlecruisers, was greatly impressed by the ranges at which his ships managed to hit. *Lion* claimed excellent fire control, straddling her targets within five to ten minutes, and doing far better than the Germans. She opened fire on *Blücher*, for example, at 21,000 yards and hit at 19,000. The German battlecruisers *Seydlitz* and *Derfflinger* were hit at 17,000 yards or beyond. Beatty concluded that hitting was possible if fire opened at any range up to 22,000 yards, far beyond anything imagined pre-war. Previous orders had envisaged opening fire at 15,000 yards. Sights and other fire-control instruments should be marked for ranges up to 25,000 yards. Plotting and range transmission had to be modified to suit; for example, the automatic range transmitter on the main (Argo) rangefinder was limited to 16,000 yards, and plotting arrangements were similarly limited. Sights in *Princess Royal* (and probably the others) were calibrated out to 20,200 yards (elevation fifteen degrees twenty-one minutes), even thought the guns could elevate to twenty degrees.

Both *Lion* and *Princess Royal* apparently used a version of rangefinder control, opening with single guns to test the range. They did not shift to salvoes until they had crossed the target. Spotting became the primary means of getting onto and holding a target because rangefinders were difficult to use. This problem and German zigzagging made time-range plotting (for rate) virtually useless; the transmitting station officer in HMS *Lion* remarked that 'on the whole I do not consider that the Dreyer Table has justified its existence.' Against a violently manoeuvring target, corrections had to be much bolder: *Princess Royal* reported needing a 1000-yard up-correction to hit at 16,000 yards. The printed report of HMS *Lion*'s experience suggested that standard corrections for all (long) ranges should be down 1000 yards (for cases in which all shots were over) and up 200 yards (to edge up to the target when all shots were short). British gunners could not distinguish the explosions of HE (high explosive) hits from German gun flashes; as they had been told, all they ever saw were shorts. If a target zigzagged violently, it would not remain at a particular range for long. HMS *Lion* suggested firing double salvoes instead of singles, what would later be called 'ladders' (see above, page 105), in conjunction with director control. This was later cited as the germ of the wartime gunnery revolution.

Although initially directors were seen as a capital-ship improvement for long-range fire, during World War I the Royal Navy realised that they were essential to cruisers and even destroyers. HMS *Dunedin* shows her director, on a mast bracket under her fire-control top, at Honolulu, 11 February 1927. Note the concentration dial in front of the director bracket, and the larger dial abaft the mast (its back is visible).

If it had to be accepted that the enemy would zigzag, then detecting and measuring his changes of course became vital. It seemed that the bearing plot, which was easier to maintain than the range plot, could help. In HMS *Lion* its data were passed up to the spotting top for comparison with the spotter's observations of enemy inclination. There was apparently no call as yet for a special instrument to detect enemy inclination. Inspired by the German tactics, Beatty suggested a moderate zigzag (two-point turn, return to course after three minutes) as a way of avoiding further damage once the Germans found the range.

In the stern chase that developed, gunlayers in forward turrets found themselves badly hampered by spray, and unable to distinguish parts of the German ships because of the dense smoke generated. Beatty decried the failure to open rapid independent fire as soon as the range was found. For him 'the difficulties of controlling [rapid fire] are as nothing compared to the disadvantages that ensue once the enemy's volume of shorts is greater than your own.' He blamed the failure on earlier Admiralty admonitions not to waste ammunition. Admiral Jellicoe later issued his own memorandum stressing the need to fire as much ammunition as necessary.[17] British fire slowed as gunlayers found it more difficult to spot their own splashes amid the splashes from enemy 'shorts'. Without director control, fire eventually had to stop even though all turrets were intact. The only director ship present, HMS *Tiger*, clearly enjoyed an important advantage. As after Heligoland, the official statement of lessons learned emphasised the need for rapid fire, as the enemy would be sure to take advantage of the respite offered by slower fire.

As for gun effects, there seemed to be no reason to doubt pre-war ideas of cumulative damage. For example, HMS *Lion* was hit fifteen times with little lasting effect. Turrets had been knocked out in the German battlecruisers *Seydlitz* and *Derfflinger*, but neither had been sunk. It might be most profitable to tear up the less heavily armoured parts of the target, as the Japanese had done at Tsushima. The Admiralty suggested that, given its anti-personnel effects, lyddite common shell was so useful that it should be used in equal proportions to AP (armour piercing) (see Appendix). Moreover, lyddite made much more visible explosions when it hit, so it made for much better spotting (powder common shells made the best splashes). Thus it might be wise to fire lyddite first, then AP. The Germans relied entirely on AP. 'For armour penetration the [German] shell seem very good, but in general destructive and incendiary effect they are not equal to our lyddite, which have proved so effective in every action of the war. Only one small fire was caused in our ships by German shell, whilst three out of the four German ships were heavily on fire at various times.'[18]

Jutland

The following year the British met the Germans at Jutland. There was not very much difference in gunnery performance between the fleets.[19] Visibility varied enormously. Firing opportunities were fleeting, targets emerging suddenly from the mist. Often there was time for only a few rangefinder readings. Afterwards the British sought alternative means of finding the range; by late June 1916 two ships were ranging experimentally on gun flashes. The British were surprised that the Germans made no attempt to break up the vertical lines of their pole masts, which British coincidence rangefinders could exploit. It was finally accepted that the existing 9ft rangefinder was insufficient for the long ranges now common. Every capital ship from the *Orion* class onwards would now be fitted with a 15ft rangefinder. Long range made more elaborate corrections necessary. Although the Dreyer range calculator had existed since 1907, the Battlecruiser Fleet had not used it, as its habits had been formed during peacetime practices and at relatively short-range wartime practices.[20] Now range correctors were to be considered mandatory, as corrections (particularly due to time of flight) were likely to be large – and magnified by large range rates. The British credited their directors with good performance despite the poor light of the latter stage of the battle.

As at Dogger Bank, the Germans zigzagged when hit, so reports again emphasised the need to measure

Although not part of this book, shore bombardment was an important big-gun function in both world wars; during World War I the Royal Navy built numerous big-gun monitors specifically for that purpose. They were among the earliest ships fitted with director control, which was particularly useful for indirect fire. For the same purpose they introduced the GDT techniques later important in other ships. They also pioneered air-observation techniques, since at long ranges they were generally firing indirectly. HMS *Lord Clive* was one of three monitors (armed with twin 12in guns) selected to have a single 18in gun (originally intended for the 'large light cruiser' *Furious*) in a semi-fixed cross-deck mounting added aft (the others were *General Wolfe* and *Prince Eugene*). Note the director bracketed to her tripod mast, and the bandstand atop her spotting top for a vertical anti-aircraft rangefinder. *Lord Clive* was the first ship to be converted; she underwent the structural modifications involved at Portsmouth between 5 December 1917 and 6 April 1918. Because the mounting could not train very much, a ship using it had to be anchored stem and stern against the current. The system used a second director (under and to port of the spotting top, atop which was the primary director) trained on aiming marks to seaward of the monitor. Director orders controlled gun train and elevation as well as the operation of the capstans capable of shifting the ship over a larger angle. *General Wolfe* used the gun for the first time on 28 September 1918 against a railway bridge at Sneskerke south of Ostende, managing to fire a shell every two minutes thirty-eight seconds. She fired a total of eighty-one rounds, beginning with the longest shot to date, 36,000 yards, for her first one. *Lord Clive* fired a few rounds on 14 October. Soon after that the Germans evacuated the Belgian coast. Conversion of the third ship was cancelled.

inclination. In most ships the rate officer initially estimated enemy inclination by taking two or three opinions into account, before plots gave him better information. The fleet gunnery-lessons committee did not call for inclinometers (it may not have realised what they might be), but they were soon ordered.[21] They measured the apparent shortening of the target due to its angle towards or away from a line parallel to the observer. A target coming towards the shooter or steaming away from him would show the same foreshortening, so some means was needed to resolve the ambiguity. Inclinometers were not available until late in the war, so ships experimented with various plots. There was a renewed call to develop means of spotting from aircraft, which might be able to see ships hidden by mist. Aircraft might observe the enemy's course directly, and thus solve the inclination problem.

Even though relatively few German ships were visible at any one time, it proved remarkably difficult to keep spotter, director, and Dumaresq on the same target.[22] The fleet committee on gunnery lessons called for installation of more precision target designators (Eversheds). Elliott Bros. was already producing a device that designated the target to the spotter and helped him keep his glasses on it; the committee wanted production accelerated urgently. Main and secondary batteries had to practice rapidly opening fire and quickly switching targets. That demanded quick target recognition; the committee stated that familiarity with the silhouettes of German ships was essential. An unstated point may have been that reluctance to open fire on individual targets was due at least as much to uncertainty as to their identity as to lack of initiative. The committee also wanted tests of German-style coloured recognition lights.

For the Royal Navy the final wartime major gun action was a battle between light forces on 17 November 1917.[23] The battlecruiser *Repulse* came up to support the British cruisers, and action was broken off when a German battle squadron appeared. This inconclusive battle off the German coast had nothing like the impact of Jutland or even of Dogger Bank.

Shells

The most important lesson of Jutland was not stated in the report issued to the Grand Fleet. The British shells just were not lethal enough. Even a ship hit repeatedly, such as the German battlecruiser *Seydlitz*, could survive; (one German ship, *Lutzow*, was crippled by British fire). It seemed that the German shells had been far more effective. Given that the British disasters were due to bad magazine practices, this is less apparent in retrospect. Apart from the ships destroyed by magazine explosions, none of the British ships suffered very badly. Apparently more was wrong with British magazine practices than right about German shells. The Germans considered their own shells too light. Had they the chance, they would later have armed their ships with heavier guns.

To the British Jutland proved that hits would be few and far between, so every one of them had to count. According to their postwar summary of wartime gunnery progress, 'theoretically no number of hits with old-type shells could cause destruction of the modern German battleship, whereas one hit *in a suitable position* with the new projectile will immobilise her...'[24] New shells were developed on a crash basis.[25] Initially it seemed that whatever superiority the Germans enjoyed was in the fuse rather than in the shell. Then it was decided that entirely new shells were needed. They were tested against mock-ups of the German battleship *Kaiser* and the new British battlecruiser *Hood*, then under construction. By 1918 the British thought the shell problem had been solved, so that any repeat Jutland would be fatal for the Germans. Postwar tests against the surrendered German battleship *Baden* seemed to prove the point. However, a note of scepticism may be in order. The British used their new shells during World War II. When they sank *Bismarck* and *Scharnhorst*, the damage tended to be cumulatively fatal, rather than instantly disabling. The shells certainly worked, but it is still not altogether clear whether the failure to inflict fatal or at least disabling damage with single hits was due to shell problems or to the sheer difficulty of disabling a very large well-designed ship.

CHAPTER 6
Between the Wars

HMS *Nelson* introduced the Admiralty Fire Control Table (AFCT) and the Director Control Tower (DCT) to British practice. A British manual later defined a DCT as a combination of the Director Tower (DT) with control personnel. The typical director crew comprised layer, trainer, sight-setter (or elevation/ deflection operator), and phone-man. The massive DCT is the lower of the objects atop *Nelson*'s tower bridge, which was revolutionary at the time. The taller object is a pedestal for her high-angle director, not fitted at the time of this 1927 photograph (at this time the platform carried an Evershed bearing transmitter and a high-angle rangefinder). Also not yet fitted were the DCTs for the secondary battery, which would be placed atop the tower on each side, alongside the gap between the main DCT and the pedestal. The top of the tower also carried a pair of 9ft tactical rangefinders (in effect, for situational awareness) roughly abeam the main DCT and not visible here. Further DCTs were aft. Each main DCT contained a 15ft rangefinder; each secondary unit, a 12ft rangefinder. Although not clearly visible, each of the three triple 16in turrets carried a 41ft rangefinder. The DCT included the first British gyro-stabilised sights, Type G, versions of which also equipped the Treaty cruisers. Maximum elevation was forty degrees for battleships, seventy degrees for *Kent*- and *London*-class 8in cruisers, and sixty degrees for *Dorsetshire*- and *York*-class 8in cruisers. The later light cruisers had Type J (Type H, the first versions of which was unstabilised, was for destroyers, but also HMS *Warspite*). Types G and J had back-laying gear, ie, they could be driven by the AFCT to provide feedback. Although the concept of the DCT included cross-levelling gear, according to the 1940 British director handbook the first cross-levelling equipment (Type C Mk I, using a vertical periscope and step-by-step transmission in twenty-five-yard steps) appeared in the *Perth* class. The first battleship version (Type BM) appeared in the *King George V* class (Mk II was earmarked for the abortive *Lion* class). Type D was for destroyers. M indicated magslip (synchro) transmission; the first Type CM for cruisers appeared (as Mk I, a version of C Mk II) in some 'Town'-class cruisers (*Newcastle* and *Southampton* had C Mk II) and in HMS *Aurora*. CM Mk II was in the *Fiji* class, and Mk III in the *Dido*s. The prominent 15ft rangefinders between the tower bridge and the funnel were for torpedo-battery control. Below is a remnant of the earlier fire-control system, a big armoured director hood atop the conning tower. The tower bridge and the DCT figured in the sketches for the abortive capital ships designed in 1920 and cancelled under the Washington Treaty. In effect the *Nelson*s were cut-down versions of the 1920 battlecruisers (they were called 'Washington's Cherry Trees,' after the famous story about George Washington chopping down a cherry tree and then admitting it because he 'could not tell a lie'). The *Nelson*s were the first British ships with triple turrets. They were designed so that all three guns had to be loaded together. This loading arrangement made it necessary to fire all three guns together (or in very close sequence), producing, at first, erratic spreads due to interference. The wing guns in each turret, however, were far enough apart not to cause trouble. The ships tried salvo firing, first with five guns (wing guns of A and Y turrets, centre gun of X turret) and then with the remaining four. Unfortunately the next salvo had to wait until all the guns were reloaded, greatly slowing fire. By 1930 it was accepted that the *Nelson*s would usually fire salvoes of all nine guns, the interference problem being solved with delay coils (ie, slight delays between rounds rather than waiting for a second salvo order). Unlike the 15in/42, this 16in/45 fired a somewhat light shell (2053lb AP) at relatively high velocity (2700 feet/second). By way of comparison, the cancelled Mk II gun of the World War II *Lion* class would have fired a heavier shell (2375lb APC) at lower velocity (2450 feet/second). The slightly earlier US 16in/45 of the *Colorado* class fired a 2100lb shell at 2600 feet/second (the heavier World War II shell [2240lb] was fired at 2520 feet/second).

FOR MOST OF THE INTERWAR PERIOD the Royal Navy expected Japan to be its future enemy.[1] In the event of war, it would form a Far East fleet based at Singapore. The war would be decided either by a fleet engagement in the South China Sea or (less probably) by an air attack on the Japanese fleet in harbour. Once the enemy fleet had been destroyed, Japan could be blockaded. British observation of Japanese fleet exercises showed that the Japanese expected to fight the decisive battle close to home waters, after auxiliary forces, such as submarines and naval aircraft, had whittled down the British fleet. This was much the way the Japanese planned to fight the US Navy. The British expected that the presence of their own fleet further south would present the Japanese with an intolerable threat, so that they would have to come south and fight without the benefit of equalisers, such as land-based aircraft. In that case the numerical battleship advantage the British enjoyed thanks to the Washington and London Treaties[2] might be decisive. Events in World War II did not work out this way, but thinking along these lines undoubtedly shaped interwar British tactical and technical development, including that of gunnery.

Another influence on this area of development was the experience gained by the Royal Navy in World War I. The most important wartime problem had been bringing a reluctant enemy to battle. If, like the Germans for much of the war, the enemy refused to go to sea, wartime experience showed that it was far preferable to destroy him at his base than to maintain a drawn-out blockade, during which other enemy forces, such as submarines, would be free to attack. The Royal Navy therefore gave considerable thought to harbour attacks, testing a remote-controlled semi-submersible explosive boat. Ultimately its means of harbour attack would be aircraft, and the idea worked well at Taranto in November 1940. If the enemy did go to sea, it might well try to evade engagement with the British battle fleet. Aircraft offered a way of finding and fixing – slowing – the enemy fleet. This idea was demonstrated against the German battleship *Bismarck* in May 1941.

As in World War I itself, tactical thinking in the immediate postwar period involved finding the appropriate balance between the available weapons with their very different characteristics.[3] Not surprisingly, initial postwar Battle Instructions were very similar to the late-war Grand Fleet Battle Orders: if possible, fire would be opened at maximum range, which meant 20,000 yards for 13.5in guns, in order to stay out of the danger zone of enemy torpedoes (assumed to have a range of 15,000 yards at twenty-five knots). The fleet would concentrate on the few enemy ships in the van, to make an impression on the enemy as quickly as possible. As the range closed, concentration would be reduced, fire being distributed over the entire enemy fleet. As a rule, however, the enemy van should not be allowed to come within 16,000 yards (due to the torpedo threat) unless the fleet commander decided specifically to accept the risk of torpedo hits in order to achieve decisive results (in other words, to 'engage the enemy more closely'). He would deploy his light cruisers and destroyers to make torpedo attacks at the first opportunity, preferably after the enemy was sufficiently absorbed by the gunnery battle not to be able to evade. The battleships would be organised in divisions which would manoeuvre separately while maintaining close support of each other.[4] A 1926 addition mentioned the need for flagships and heavy ships to maintain tactical plots; special plotting facilities had recently been installed in British capital ships.

The wartime injunction that night action was generally to be avoided still held, but July 1922 additions to the Battle Instructions by the Atlantic Fleet included the comment that battleships could be effective against each other at night 'if night training is good'. This was a germ of later British naval practice. The 1928 Battle Instructions, however, counselled that if the enemy had not been defeated by nightfall, action should be broken off, and the enemy fleet shadowed so that fighting could be resumed at daybreak.

Lecturing at the Royal Naval War College in 1920, Captain (later Admiral) C V Usborne pointed out that changes in gun-operating practice and in magazine protection had, it seemed, cured the devastating problems demonstrated at Jutland. As before that battle, it now seemed that it would take multiple hits to destroy a battleship. Bulging had now considerably devalued the torpedo. Usborne agreed that it was important to concentrate fire: quickly disabling a few ships would reduce enemy fire far more effectively than spreading fire along the enemy line. The positive effect on morale brought about by blowing a few ships out of the enemy formation might well be profound. He saw little point in fighting at long range, however, because hitting rates would be low; he wanted to close as quickly as possible to decisive range, which he estimated to be 14,000 yards. Usborne accepted that his own ships would be punished on their way in to such an engagement.[5]

Usborne considered the gun the key weapon because he thought the new bulges and and analogous anti-torpedo protection would enable ships to survive multiple torpedo hits with most of their capability intact. He was thus unimpressed by the fact that new torpedoes could reach 16,000 yards. Others disagreed (the British certainly continued to work hard at destroyer attack tactics, for example), but Usborne showed that the driving force of the pre-World War I period, the torpedo threat, had largely dissipated as a dominant theme in gunnery.[6]

The 1927 edition of the Battle Instructions still envisaged initial concentration fire at long range, but admitted that this might be difficult to achieve. Since it was essential to ensure quick destruction of the enemy fleet at long range or in poor visibility, it was essential to make the most of any opportunity; that in turn required an efficient means of designating targets within the fleet. It was also vital to be able to switch concentration targets quickly without disorganising the fleet. As in the past, the favoured torpedo tactic was a mass attack, with the comment that multiple attacks might succeed because enemy manouevres to evade one might place the targets in the path of another. These were the 'browning-shot' attacks imagined pre-war, but mainly by destroyers rather than battleships. Opportunities would be fleeting, perhaps six minutes of torpedo firing time in any one set of attacks.

Although the next edition of the instructions retained the idea of an initial long-range concentration, the emphasis shifted to a fighting range of 12,000 to 16,000 yards, which was much what Osbourne had in mind.[7] It was accepted that the fleet would have to fight within enemy torpedo range, but that this could be neutralised by unexpected British manoeuvres. Moreover, the enemy battleships would probably find it difficult to fire torpedoes because they would be manoeuvering to evade the greater torpedo threat presented by the larger number of British battleships. Overall, the tactical objective was to produce the maximum volume of fire, from both gun and torpedo, before the enemy could.

The attitude towards night combat evolved. The 1931 Battle Instructions observed that night action would generally be to the advantage of the weaker fleet (because a night battle would be a mêlée), but circumstances might require it if the enemy showed signs of defeat or confusion, or if he could not be defeated before nightfall. By way of contrast, the 1934 edition stated that under 'certain conditions' night actions between capital ships would be sought. An insert went further: 'night action between heavy ships…must be regarded as a definite part of our policy, to be taken advantage of when circumstances require…' The context was generally a day action that had not ended conclusively by nightfall. This was exactly what happened at Matapan seven years later. The final pre-World War II set of instructions, the 1939 Fighting Instructions, also carry the 12,000 to 16,000 yard prescription, with the curious observation that it was associated with the 'superior fighting qualities' of the Royal Navy. Probably the key to better night performance was the new mechanised plotting, which made it possible for a commander to keep track of an evolving tactical situation (situational awareness in modern parlance). The need to maintain a night plot may explain interest in non-visual rangefinding methods, such as correlating radio and acoustic signals. They would make it easier to maintain an accurate plot of own-ship positions at night, which in turn would make it less likely that ships would engage friendly ships.[8] In the 1930s the Royal Navy was unique in developing tactics and techniques for a night fleet engagement including battleship gunfire (other navies generally relied on torpedo fire, supported to some extent by cruiser gunfire).⁹ This capability was displayed at Matapan in March 1941, an Italian cruiser division being wiped out by fire from the battleships *Warspite*, *Valiant* and *Barham*.

Plotting made divisional tactics possible. The fleet's reaction to Jutland, said one officer, was to make decentralisation 'a fetish'.¹⁰ The fleet remaining after the treaties made some form of divisional organisation almost inevitable, as it contained three battlecruisers, seven fast battleships (*Nelson* and *Queen Elizabeth* classes), and five slow ones (*R* class), which generally occupied a flanking position. The battle force was generally organised in two divisions, far enough apart to give them freedom of action in the face of torpedoes and to preclude the sort of 'browning shots' so feared before 1914.

By the mid-1920s British battleship designers working on projected ships were providing sufficient armour for them to fight at a minimum range of 12,000 yards, as was evident in a sketch design prepared before the abortive 1927 Geneva Conference on further naval arms limitation.¹¹ At that time it seemed that new ships might be laid down in 1931 (under the agreement reached at Washington in 1921), so battleship staff requirements were developed in 1928. Because the attempt to limit guns to 12in had failed, the standard of protection was set at 16in, if possible. In the new treaty signed in 1930 the horizon for new construction was pushed back again to 1936, so work on basic requirements resumed in 1933.

By 1929 British naval intelligence was reporting that 'certain foreign navies' were practicing at 30,000 yards, far beyond British capability.¹² If it encountered such a fleet bent on action, the British fleet would have to close to its own preferred range of 20,000 yards or less. Without experience of firing at 28,000 yards or beyond, the British could only speculate on the damage they would have to accept while closing with the enemy. Long-range gunnery tests were assigned to *Nelson*, *Rodney* and *Hood* specifically to learn what could be done. The British considered their fleet somewhat faster than the enemy's, but the difference was small. A tactical table exercise showed that it would take fifty minutes for a twenty-knot force to close with a nineteen-knot force, reducing range from 34,000 to 20,000 yards. At long ranges, moreover, British battleship magazines could be penetrated by high-velocity 14in shells (as in the US and Japanese fleets) at 19,000 yards; (*Barham*, the first *Queen Elizabeth* to be rebuilt with added deck armour, was considered immune to 29,000 yards). The Tactical School estimated that one British battleship would be disabled or sunk for every four enemy battleships firing during the run-in.¹³

When work on battleship Staff Requirements resumed in 1933, the US long-range capability was considered the single most important new gunnery development to date. It made deck protection much more important, as did improvements in bombing, particularly dive bombing with 1000lb AP (armour-piercing) bombs, which the US Navy had also demonstrated.¹⁴ Notes on the initial designs prepared in 1934 show sides proof at 12,000 yards, but only indicate the resistance of deck armour to various bombs.¹⁵ However, a February 1936 staff memorandum stresses the need to give the ship sufficient protection to enable her captain to bring her into decisive range, ie, to survive enemy long-range fire for long enough to reach the desired battle range, below 16,000 yards ('to make the most of the national characteristics of our personnel'). That justified a thick protective deck atop the belt.¹⁶

Thus the new *King George V*-class battleships were designed to fight inside 16,000 yards. They therefore devoted more weight to armour than their US equivalents, the *North Carolina*s.¹⁷ The selection of around 15,000 yards as the inner edge of the immune zone suggests that this was still the expected decisive range when the ships were designed in 1935–36. The emphasis on fighting at about 15,000 yards made it acceptable to install new fire controls atop tower bridges rather than, as in other navies, atop tower masts.

For a long time to come, however, most of the British battlefleet would consist of older ships. In 1932–33 the Admiralty laid out a schedule to rearmour some of these vessels between 1934 and 1940.¹⁸ Much had been learned about deck and underwater protection from recent trials against the battleships *Empress*

In contrast with its US and Japanese rivals, the interwar Royal Navy invested only limited funds in battleship reconstruction, and until the last series of ships it showed no interest in installing the new computer fire-control systems. In September 1933, when it was considering modernisation, the Board of Admiralty was told that, in current terms, the United States had already spent £16.3 million on modernisation (and planned to spend another £15.8 million), the Japanese had spent £9.3 million – and the Royal Navy had spent only £3 million. The Royal Navy concentrated on protection against air attack, which included deck armour expected to enable ships to survive longer-range shellfire. HMS *Barham* was a transitional case. Modernised between January 1931 and January 1934, she was the last in her class to have her funnels trunked together. She was the first in her class to receive additional deck armour but was not re-engined, nor did she receive new fire controls. She retained the big armoured rangefinder hood atop her conning tower, and the director (in a thimble canopy) below her spotting top. B turret shows her 30ft rangefinder (as in X turret; A and Y turrets had 15ft rangefinders, which did not extend beyond the sides of the turret). Her compass platform was enlarged, roofed, and fitted with a flared steel screen (she had had a windscreen for some time). Above it was built a new enclosed torpedo-control position (the Royal Navy was unusual in retaining torpedo tubes in some battleships) with its own rangefinder in a hood. Similar enclosed (or open) platforms had been installed earlier. The platform above that carried the 15in director. Air-defence improvements included the two platforms built out from the spotting top for pompom directors, and high-angle directors on the spotting top and aft (the spotting-top director has not yet been fitted; it was in place by July 1934). Because air attack would include strafing attacks on the ship's superstructure, the response to the air threat included bullet-proof plating for the compass platform and the spotting top. Note also the MF/DF loop atop the spotting top. HMS *Malaya*, which had already had her funnels trunked together, received a similar refit in 1934–36, including the addition of a cross-deck catapult (*Barham*'s was atop her Y turret). She emerged with a different profile, and without any torpedo-control position (her tubes had been removed).

of India and Marlborough. Although Empress of India had flooded after a hit by a diving shell, it seemed that existing underwater protection would protect newer ships. The main problem was deck armour, to protect magazines at long range and to protect engine rooms, which presented a large deck area, against bombs. At this stage nothing was said about gun range or fire-control performance, at least not in papers written for the Board of Admiralty. It was repeatedly pointed out that British spending on capital-ship modernisation significantly trailed that of the United States and Japan. Although war was not likely in the near term, matters might well worsen in the Pacific, in which case Britain should not have to build new battleships (to match Japan) while also modernising existing ships.[19]

A survey of the British battlefleet produced a desired standard of protection, to some extent a reaction to the need to deal with longer enemy gun ranges. It turned out that the 15in battleships and HMS Repulse all needed extra armour (Renown had been given considerable extra deck armour in 1926); Hood met the desired standard. On this basis HMS Barham was taken in hand in 1932 and Repulse in 1934; there was debate as to whether any of the R class should receive extra armour, given that it had already been decided to scrap them before the Queen Elizabeths. As the international situation darkened, it seemed that at least two of them would be modernised, and there was some question as to whether all the Queen Elizabeths were worth doing. In 1933 it seemed that this modest reconstruction would embrace the Queen Elizabeths, Royal Oak, and Revenge. Re-examination of the state of the ships caused Malaya to replace Revenge in the 1934 programme; the R class never got the extra deck armour.

These were still moderate refits, and they did not entail any major improvement in main-battery capability. In 1934, however, re-examination of the long-term construction schedule reminded the Admiralty that some existing ships might be required to serve as long as thirty-three years. The Board now focused on the remaining Queen Elizabeths and the battlecruiser Renown. Perhaps they should be re-engined to keep them reliable. Initially that would have meant Malaya, but Warspite was already having engine trouble, so she was chosen. Re-engining involved so much work that it was worthwhile to re-examine other possibilities for improvement.[20] Only at this point was main-battery modification considered, at least at the Admiralty Board level. Board approval came in stages: first increased gun elevation, as the American and Japanese navies had already done (in this case, to gain 5500 yards, according to a 1933 analysis); then new 6crh (crh means calibre radius head; a higher number means a pointier shell) shells (2200 yards, also advocated in 1933); and then new fire-control tables (AFCT Mk VII), design work on which DNO reported in 1934. At this stage accounts of planned reconstruction were still accompanied by the phrase 'minimum modification to bridges', although the conning tower was to be removed. Ultimately the bridge was completely rebuilt, among other things to accommodate the DCT (Director Control Tower), which worked with the AFCT.

Warspite thus became the first British capital ship to undergo full main-battery modernisation. Malaya and Barham already having received limited modernisation (including increased deck armour), only three Queen Elizabeths were subject to full modernisation. Of the three battlecruisers, Renown was selected as the first to be modernised; Repulse and Hood to have followed. Hood was to have followed Queen Elizabeth, but she was caught by the Munich crisis and then by the outbreak of war. She could not be taken out of service long enough for reconstruction (and, as it turned out ironically, her deck protection was considered good enough). The main argument for rebuilding her was that she was expected to serve until 1953, which was impossible without re-engining. In this context the First Sea Lord wrote that it would be a matter of 'eternal regret' if the rest of a proposed reconstruction was not carried out.[21] The R-class battleships were not considered worth rebuilding, and by 1939 they were scheduled for disposal as soon as the new King George V class entered service.

Contemporary Admiralty documents show no concern with fire-control modernisation. The two vital issues were machinery and horizontal protection, the

Given limited funds, the Royal Navy chose to maintain the large cruiser force it thought necessary to protect global British shipping, and also to build new destroyers and submarines. From a surface-battle point of view, the existing battlefleet seemed adequate once ships had been bulged (to resist torpedo attack) and magazines had been modified (most importantly, after dangerous pre-Jutland practices had been abandoned). What money there was, went mainly into exploiting air observation (to improve fleet gunnery performance) and into air defence. HMS *Ramilles* is shown after a 1933–34 refit, with the new high-angle directors (for the HACS Mk I system) atop her fighting top (in this case replacing a high-angle rangefinder) and bracketed to her tripod mainmast (newly converted from a pole by adding struts, to provide support for the new director). Multiple 2pdr pompoms have been added on platforms abeam her funnel, replacing the searchlights previously mounted there. Also visible are the quadruple 0.5in machine guns added abeam her conning tower. In addition, pompom control platforms have been added on each side of the 15in director bracketed below the spotting top. At the same time a catapult was mounted atop X turret, the earlier flying-off platforms being landed (the after control position was landed). Note the tactical rangefinder above the bridge, usually described as a torpedo control unit, but essential for plotting. Even so, it was replaced by a large open platform the following year. During this refit the after pair of torpedo tubes was removed, the remaining ones going by June 1938. The concentration dial forward had been relocated to the mainmast (where it is barely visible) during a 1923–24 refit; it had been moved from the face of the control top to a pole above the control top during a 1919–21 refit. The upper bridge had been enclosed during a 1926–27 refit, but later it was given bullet-proof plating to protect against strafing aircraft. *Ramilles* and other R-class battleships seemed a reasonable alternative to the more expensive *Queen Elizabeth*s, but ultimately they were deemed less valuable because they were slower. They also had considerably less internal space for new features such as a modern analogue computer, but it is not clear from surviving documents how much this limitation affected decisions about rebuilding them. They were fitted with Admiralty Fire-Control Clocks Mk IV to control their secondary batteries. HACS Mk I also demanded internal space. By 1939 the R class were considered second-line units awaiting scrapping once the *King George V* class appeared. Even so, *Royal Oak* was given 900 tons of additional deck armour (4in over magazines and 2½in over engines over the existing 1in plating) during a major refit begun in June 1934. In February 1939 similar work was proposed for *Royal Sovereign* and *Ramilles*, and bridge modifications were also suggested; gun elevation would have been increased to thirty degrees. The work was delayed by the need to refit *Nelson* and *Rodney*, and then stopped by the war; but the proposal suggests that these ships were still considered valuable. The additional deck armour would have been comparable to that in *Barham* and in the heavily rebuilt *Queen Elizabeth*s. Although nominally also part of the programme to improve survivability against air attack, it and higher-elevation guns would have given the ships greater effective gunnery range. Gun elevation was never increased, but three ships did receive 2in deck armour over their magazines in 1942 (*Resolution*, *Royal Sovereign*, and partially in *Ramilles*).

latter generally a matter of protection against bombs, not long-range plunging fire. It might seem that only specialist gunners (and DNO and DGD) understood the difference between existing fire-control systems and what ships could have. Because the printed minutes of Admiralty Board meetings are so brief, and because so little of the internal Admiralty correspondence has survived, we cannot tell whether the gunnery experts made their case to the board, or whether the board simply accepted that gunnery modernisation ought to take place once so much other work was being done. Once it had been decided that ships should be re-engined, space became available for installation of the massive AFCT. In retrospect the decision to re-engine and the decision to fit the new fire control seem to fit together perfectly, but that may not have been so obvious at the time. Probably the *R* class had too little internal space, but that is not obvious from the documents (and in 1940 a proposal to fit a surplus AFCT on board HMS *Resolution* was approved).

Unlike its foreign rivals, the Royal Navy made no effort until the late 1930s to increase gun range because it did not consider long range very effective. The question was not how to extend British gun range, but rather how to survive under longer-range enemy fire while closing to decisive range. That usually meant either negating enemy long-range fire or extending the outer edge of the immune zone so that the enemy fire would be ineffective at worst. New British ships could fire at long ranges, but those ranges did not figure in Admiralty internal discussions of the virtues of the new ships. This explains the interest in deck armour before any interest in increased gun elevation or in a new AFCT for rebuilt battleships. The British were certainly aware that their situation was deteriorating: in 1937 their naval intelligence reported that the Japanese planned to open fire, not at the previously reported 22,000 or 23,000 yards, but at 30,000 yards. And even this was actually a gross underestimate. They were also unaware that the Japanese had modified their ships for much higher speeds; the Queen Elizabeths and Nelsons were no longer at least as fast as any of the Japanese battleships.[22]

Given their assumptions, the British were quite willing to adopt measures that would negate long-range fire by both potential enemy fleets. Long-range fire required air spotting, so the Royal Navy tested radio jamming (which proved inadequate in the Mediterranean Fleet) and considered using fighters to drive off enemy spotters. It also became interested in torpedo attack as a way of disrupting enemy fire. To do that at long range, it planned to use torpedo bombers (destroyers might attack in combination with them). Smoke could prevent surface spotting (the British considered laying smoke *over* the battle area to negate air spotting). A suggestion that the two longest-range ships, Nelson and Rodney, be detached to provide supporting fire while the rest of the fleet closed in was rejected for fear that the enemy would simply concentrate on and destroy those two ships.

British pre-war thinking about decisive range helps explain why HMS Hood ran in towards the German battleship Bismarck in May 1941. Her tactic is usually ascribed to her commander's fear that her decks could be penetrated. He certainly had that problem, but he was probably running in to reach what the British still considered effective range. The ship was lost, it seems, not to some fundamental flaw – and not to the kind of outrageous magazine practices which had cost the battlecruisers at Jutland. She suffered an unlucky hit that was effective because she had been modified between the wars on a piecemeal basis.[23]

Air assistance was a vital new feature of the fire-control picture. The Royal Navy began with air observation, in which an observer aloft (preferably directly over the firing ship) reported target course and, somewhat less precisely, speed. These data could be inputted directly into the Dumaresq of a Dreyer Table. Air observation of enemy course and speed was simpler than full air spotting, which required continuous contact between aircraft and firing ship. Royal Navy experiments with air observation began during World War I, but full air spotting probably became possible only after interference between engines and radios was eliminated in the late 1920s. Air observation may explain why the Royal Navy accepted lower director positions than those of other major navies, and a much shorter (15ft) main rangefinder.[24]

Data from aircraft came directly into a ship's transmitting station, not via the control position aloft. To what extent should the transmitting station take over control of the ship's armament? When HMS Warspite was rebuilt in 1934–37 with a new synthetic fire-control system (AFCT Mk VII), she used 'plot spotting' whenever aircraft were available. Otherwise her control officer still relied on his own observations, plus those of three outlying spotters. Given an effective computer, he no longer had to observe the target continuously. The fleet sought an appropriate balance between full control from the transmitting station and rigid application of all spotting rules by the control officer. By the late 1930s spotting rules themselves were more complicated, and many ships found that it helped for the transmitting station to order the details of strings of salvoes.[25] At the outbreak of World War II the Royal Navy still had not decided how control should be shared between the transmitting station

(particularly if it had a synthetic calculator such as an AFCT) and the control officer aloft.

The Royal Navy probably felt that it could rely on air spotting because it would be fighting in clear Pacific weather. The reality, that it fought mainly in bad European weather, must have been a very unpleasant surprise. Royal Navy gunnery publications of the interwar period are filled with experiments with and then the applications of air spotting, which made very long-range gunnery quite practical.

Rethinking fire control

The postwar generation of fire-control systems can be traced to a 1918 Grand Fleet committee called to review the design of the Mk V Dreyer Table planned for HMS *Hood*.[26] The fleet wanted a range-keeper that could project ahead own and target motion, and some means of deriving target inclination from a plot. The existing table did neither. It was valued mainly because it brought together all relevant data. However, the plot was hidden from view by a mass of overhead fittings, such as the rate grid, the Dumaresq, and the carriage for the pencil. The clock (integrator) was becoming overloaded with plots and fittings.

Gunners wanted it to show not only estimated rangefinder range (from the clock), but also ordered gun range, for comparison with spots. However, the fire-control system was based on rangefinder range. The connection between gun and rangefinder range was the Dreyer calculator, which had to be reset to reflect a changing range rate. Correction of the range rate therefore entailed mental arithmetic, hence delay and possibly human error. Moreover, the Dreyer calculator could not keep up with high range rates. For example, every time the ship changed course a new rate had to be read off the Dumaresq or rate receiver and a new wind read off and set; the total correction was set on a dial which moved the zero of the spotting corrector. Enemy course and rate were similarly handled. The importance of such corrections had recently been demonstrated in high-speed throw-off firings (ie, with high rates: the target would travel a considerable distance while the shell was in flight). With the new

The Kent-class 'Treaty' cruisers were laid down before the DCT was ready. They received AFCTs (Mk II) but had earlier-style topside arrangements. Initial plans showed a tower topped by separate director and 12ft rangefinder. As illustrated by HMS Berwick, ships were completed with director and rangefinder combined, as in battleship armoured rangefinder hoods, with a separate windowed control position below, housing gunnery officer and spotters. The level below also housed the plotting position, vital for tactical command and control. The computer was in the transmitting station below the waterline. In the 8in/50 mount, note the sight visible on the side of the gun-house forward of and below the 'ear' of the rangefinder. The rated rate of fire was eight rounds per minute. Even though the two guns of cruisers with twin turrets were well separated, they (like the Nelsons) suffered interference, and ultimately all the 8in and 6in cruisers were fitted with delay coils. In 1926, in response to more heavily armed US and Japanese 'Treaty' cruisers, the British considered adopting triple turrets, which they expected would probably fire about a quarter more slowly: about eighteen rounds per minute. DNO commented at the time that it was more difficult to design a triple 8in than a triple 16in turret (as in the Nelsons) because the turret would be far smaller but the men inside would not. A proposal to contract for triple-turret designs under the 1927 budget was dropped in favour of collecting intelligence about US experience with the triple mounts in the new Pensacola class then under construction.

interest in concentration of fire, ships needed a second plot dedicated to that purpose, showing concentration range and mean rangefinder ranges. The future Mk V table should accept and display data from three outside sources, one of which might be an airplane. Improvements wanted by the fleet included the straight-line plot described in chapter 2, page 54.

The 7 February 1919 committee report concluded that the Dreyer Table, however modified, was obsolete, its basic principles discarded: a complete redesign was essential. The future table would separate the geometric and ballistic sides of the problem (thus making it easier to change ballistics). It would comprise a clock generating the gun range, deflection, and bearing, and a plot comparing clock output with observation (including spots). For clarity, it would display own and target motion separately, rather than together, as in a Dumaresq; for example, unlike existing tables, this one would display own and target contributions to deflection. Each component of the difference between rangefinder and gun range would be shown separately so that it could be evaluated individually. Plots (gun and rangefinder range, bearing, and inclination) would all be of the straight-line type. Own-ship data would be entered automatically. The table would transmit gun range to both a gun-range plot and to the guns themselves.

To unclutter the table, some functions would be carried out on remote dummy plots. Their outputs could be projected onto the table from overhead, operators entering them using follow-ups. This device, for rate grids, was already on board two ships (it was being perfected by Elliott Bros).

The table would be adapted to indirect and to concentration firing (eg, it would automatically plot the gun range of a consort ship). Integral with the clock would be a device automatically correcting consort data for position in line (PIL). Ultimately the table would make it possible to integrate a ship into a group, just as director firing integrated each turret into a ship. Table configuration would be dictated in part by the requirements of aircraft observation, the key to really long ranges.

The committee proposed combining the best features of the Dreyer Table and the Argo Clock with those of the US Ford Range-keeper, which the Royal Navy had encountered during the the First World War. It recommended that a new Fire-Control Committee be formed comprising the senior engineers responsible for the Dreyer Table (G K B Elphinstone Esq, of Elliott Bros) and for the Argo Clock (Lieutenant Commander Isherwood, formerly of Argo, and later in the Mine Department), the Royal Navy officer responsible for maintaining the tables and clocks during the war (Lieutenant Hugh Clausen RNVR, then of HMS *Benbow*), and a gunner to provide an operational point of view (Lieutenant S Dove RN of HMS *Royal Sovereign*). DNO rejected Isherwood, unless he became an Admiralty employee (presumably to maintain secrecy). He and his old assistant from Argo, D H Landstad were hired to design the new table. Clausen served on the committee as a civilian.[27] The committee produced the Mk I and II Admiralty Fire-Control Tables (AFCTs). When it was dissolved, Clausen became senior British fire-control system designer.

The committee envisaged three connected system elements: the table, a new type of director (the DCT [Director Control Tower] described below), and a new kind of transmitter. The combination was revolutionary. Design and construction of prototypes was slower than expected. Remarkably, even though scheduling precluded sea trials, AFCT Mk I was installed on board the new battleships of the *Nelson* class.[28] Two test DCTs were made by Vickers, a large one for battleship trials and a smaller one for the cruiser *Enterprise*. Given DNO's insistence that DCT and table were integral, it is surprising that the cruiser was fitted not with a prototype table but rather with a Dreyer Mk III*.

The AFCT Mk I that was installed on the *Nelson* class was, like the Pollen clock, a synthetic system. Construction was modular, with a total of twenty units connected by shafting under a false floor.[29] In addition to the table in the main compartment, the system used a dummy-plot compartment, a gun gyro room (which also contained elements of the transmitters), and a junction-box compartment. The gyro room was intended for the master gyro and constrained gyros used for remote power control (in fact these ships never had master gyros). Because it was difficult to ensure that electric motors would run at exactly the desired speeds, the table was driven mainly by a series of multi-cylinder air motors driven by an electric compresser.[30] The linear arrangement of the table made supervision relatively easy for the table officer.[31] Operation was automated as far as possible, with manual adjustments of data.

Every time enemy settings changed, plotting began on a new basis. It was recommended that suggestions (estimates of enemy course and speed) be obtained every half minute where possible, and a pencil line drawn across the plot each time one was made. Presumably targets typically changed course on this sort of time scale. The last half-minute of plot data would be taken most seriously, but information back to the last line across the plot would also be taken into account. When inclination was near ninety degrees,

AFCT Mk IX was designed for the new World War II battleships. It equipped the *King George V* class and would have equipped the abortive *Lion* class. The success of HMS *Prince of Wales*, with a raw gunnery crew, against the German battleship *Bismarck* can be attributed in large part to the success of this table. Less automated techniques, such as those employed by HMS *Hood*, required far more training to produce a highly coordinated fire-control crew. This idea that automation was key to success in snap actions can be found among Royal Navy officers supporting Pollen's form of automation about 1911. They suggested that he adopt Kodak's slogan, 'you press the button, we do the rest', which had originally meant that the photographer pressed the button, but Kodak developed the photograph. In gunnery it was intended to mean that the emphasis should shift from making the system work to selecting the targets. The complexity of the table shows that even an automated system needed a lot more than a simple button; but it was still a lot better than its predecessors. Mk IX had the same U-shape as in the rebuilt battleships because it made for better supervision by an officer inside the U. The overhead elements of the earlier tables were finally eliminated, and both range and bearing plots were provided in addition to the enemy rate plot (on a separate unit). The cross-wires of the enemy-ship dial (D18) were used for cross-cuts to estimate enemy course and speed. The dummy data unit to the right was intended to reduce the clutter on the main computer. Note the indirect spotting dial (D15) intended to support shore bombardment (indirect fire). The Royal Navy practiced shore bombardment throughout the interwar period, although it seems not to have adopted US-style high-capacity shells until World War II.
(A D BAKER III)

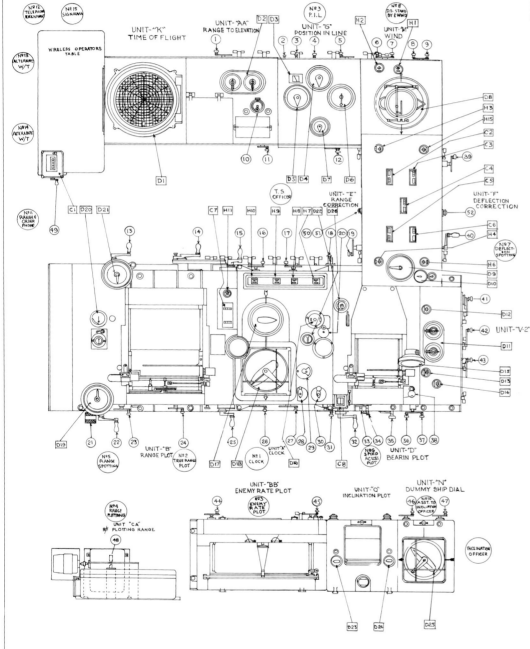

Reference To Handles
No. Function
1 Gun Elevation
2 Datum Distance
3 True Range Follow Up
4 Range P.I.L.
5 Datum Angle
6 Wind Speed Across
7 True Wind Speed
8 True Wind Compass Direction
9 Wind Speed Along
10 Dip Setting
11 Target Height Correction
12 Bearing P.I.L.
13 Range Spotting
14 Clock Range
15 Range Tuning
16 Own Speed Along
17 Own Speed Across
18 Timing Clutch
19 Own Wind & Ballistic Corr.
20 Own Speed Setting
21 Range Finder Clutch
22 Range Spotting
23 Range Finder Tuning
24 Suggested Speed Along
25 Range Tuning
26 Enemy Speed
27 Enemy Course

No. Function
28 Enemy Range Corr. Applied
29 Slip Angle
30 Enemy Deflection Applied
31 Clock Setting (Emerg.)
32 Training
33 Tune Train Clutch
34 Cloci Setting Clutch (Emerg.)
35 Direct-Indirect C.O.S.
36 Suggested Speed Across
37 Indirect Line Spotting
38 Re-Centre Pen
39 Own Wind & Drift
40 Spotting
41 Compass
42 Cross Leveling
43 Bearing
44 Shift Plot
45 Range Finder Tuning
46 Enemy Course
47 Enemy Speed
48 Range Plotting
49 Enemy Course Aircraft Report
50 Thermometer
51 Barometer
52 Drift Setting

Reference To Hunters
No. Function
H1 Wind Speed Along
H2 Wind Speed Across
H3 Enemy Range Correction
H4 Own Wind & Ballistic Corr.
H5 Won Wind & Drift
H6 Enemy Deflection
H7 Enemy Speed Across
H8 Own Speed Across
H9 Enemy Speed Along
H10 Own Speed Along
H11 Clock Range

Reference To Counters
No. Function
C1 Enemy Course Aircraft Report
C2 Calculated Enemy Range Corr.
C3 Own Wind & Drift Deflection
C4 Datum Deflection
C5 Own Wind & Ballistic Corr.
C6 Calculated Enemy Deflection
C7 Clock Range Sensitive & Power
C8 Enemy Compass Bearing

Reference To Dials
No. Function
D1 Aircraft Spotting
D2 Director Setting
D3 Datum Distance
D4 Range P.I.l.
D5 True Range Follow Up
D6 Datum Angle
D7 Nearing P.I.l.
D8 Wind Resolvng
D9 Deflection Spotting
D10 Spotting & Rejection
D11 Cross Leveling
D12 Own Course ⎫
D13 Bearing   ⎬ Magslip Transmitter Indicators
D14 Distance Across ⎭
D15 Indirect Spotting
D16 Own Speed
D17 Own Ship
D18 Enemy Ship
D19 Range Spotting
D20 Firing Sector & Watch
D21 Range Spotting
D22 Table Time
D23 Inclinometer D.C.T.
D24 Inclinometer Aft
D25 Dummy Enemy Ship
D26 Timing Watch

salvoes would straddle. Shifting to rapid broadsides as soon as a salvo straddled would almost inevitably lose the target, and a regaining ladder would have been needed.

New rules were finally promulgated on the eve of war, in 1939.[51] They established four standard groups: a deflection group (to find or regain line), the old ladder (to find or regain range), a zigzag group and a rapid group (to increase hitting once range and rate had been verified). All groups comprised two or three salvoes, except for the rapid group, which might have four. The third salvo could be withheld if the first two showed it would be wasted. With deflection and zigzag, the correction for the third in the group was to be applied whether it was fired or not. In the rapid group, the next salvo should be ordered not later than the fall of the second salvo, provided the target was held, in which case firing rapid salvoes as groups should not reduce the volume of fire as compared with simple continuous rapid salvoes.

The idea was generally to outline the area in which the target should be, then fire a third salvo at the expected target position. Thus the deflection group consisted of one salvo to the right, the second to the left, and the third at the expected bearing. Ladder steps were 400 yards, zigzag 200 yards. For capital ships the zigzag spacing was 100 yards. Rapid groups were three or four salvoes fired as quickly as possible at the deduced hitting range, without spreads (three groups for a modern destroyer). Typically a ship would try to find line first (firing a deflection group), then range (a ladder). If the deflection group fell short, the ladder would begin with a 400-yard correction if the shooter was confident of the range, or an 800-yard correction if the range was in doubt. Aircraft reports could form the basis for larger corrections, since the aloft observer could estimate range errors (and could see overs). A target should be considered lost when a zigzag failed to enclose it (no straddles and no salvoes on either side) or when two successive salvoes of a rapid group

At Port of Spain on 14 March 1950, HMS *Glasgow* shows her cruiser DCT, which carried a 15ft rangefinder across its rear and had, in this case, the 'cheese' antenna of Type 274 radar on top. The windows on its face indicate the operators within. The two circular shields covered Kent clear-view screens for the gyro-stabilised (P) sights for the layer and trainer (right and left in this view). The two horizontal windows were for the spotting officer and the rate officer (left and right), who had to slew their binoculars to correct for fire-control errors. The small paired windows under and to either side of each slot are for the spotting glasses of these two officers (one of the windows of the left pair is not visible). Another pair of windows, above these sets, is for the control officer (who had another wide window, not visible here). At top right is a hatch covering the opening for the inclinometer.

The standard British World War II large-cruiser fire-control computer, AFCT Mk VI, is shown on board the museum ship HMS *Belfast*. A plate on the table indicates that it was serial number 18, made by Elliott Bros. A separate AFCC Mk VI (in another compartment) controlled after turret or turrets in divided fire.
(AUTHOR)

fell on the same side. When the target was lost, a ship would typically fire a ladder (beginning with the range on the sights) and then a zigzag. Other combinations were used if the target was lost for line. There were many other detailed rules, but these give some idea of the sophistication of the process and the detail with which it had been thought out, based on fleet-wide experimentation.

Concentration fire

It turned out that the spread for an entire battle division could be held to a reasonable level, so concentration firing improved the overall hitting rate, beyond that which individual ships could achieve. Concentrating fleet firepower on part of an enemy's line would impose the highest rate of damage. Two alternatives were developed, individual ship (GIS, originally called ACY for the signal used) and master ship (GMS, originally MSC).[52] GMS offered higher hitting rates, and by 1923 had been chosen as primary technique. It required intense training and also that all ships involved had compatible fire-control systems and guns with similar ballistics. When HMS *Warspite* recommissioned in 1937 with the first new-generation battleship system (AFCT Mk VII) it turned out that she could not combine in GMS fire with ships still using Dreyer Tables.[53] In GIS, ships spotted their own salvoes but fired together at the same target. As during the First World War, ships distinguished their splashes by firing only during a fixed time sector, typically fifteen seconds per minute for a four-ship concentration. That made it difficult for ships to achieve full output. Even with only two ships firing (thirty-second sectors) it took excellent gun drill to achieve the maximum of two salvoes per minute per ship. Moreover, under action conditions, with the target obscured by smoke, it might be difficult if not impossible for each ship to spot salvos properly.

Concentration required communication between ships, and it seemed to work much better with air observation. Much effort went into finding a better (and less vulnerable) replacement for the wartime Type 31 concentration radio set. Its antennas could be shot away. For a time in the 1920s the Royal Navy experimented with a Towed Electrode Method (TEM) of inter-ship communication (as well as with visual indicators and masthead lights). TEM was too confidential to be described in detail in surviving publications, but it seems to have used what is now called Underwater Electric Potential (UEP). Ultimately the British adopted VHF radio for coordination.

After trials it was discovered that the *Nelson* table was not suited to master-ship concentration fire; the ships were always limited to individual concentration tactics. One reason for this was that it was very difficult to distinguish among the nine-gun salvoes each ship had to fire. At very long range, moreover, the salvo from one ship might still be in the air when the other was ready to fire. The companion cruiser table (Mk II) was tested successfully in single-ship firing (1929), but it was not modified for master-ship operation so as not to interrupt the completion of the *Kent* class. The *London*-class tables were modified for master-ship fire.[54] The modification was extended to the earlier *Kent* class but proved unsuccessful in the *Nelson*s.

CHAPTER 7
The Second World War

HMS *Nelson* is shown on 4 July 1944. The director visible just abaft the mainmast is the after main-battery DCT. Note that it had no ranging radar; British radars were in short supply.

THE GREAT WARTIME TECHNOLOGICAL SURPRISE was radar, which finally made it possible to fight at long range at night – as HMS *Duke of York* showed in December 1943 when she sank the German *Scharnhorst* while suffering virtually no damage in return. First-generation metric wavelength radars (eg, Types 79, 279, and 281 in battleships and cruisers) were intended primarily for air search, but they could detect surface targets beyond the horizon, using surface waves. Thus, in an action on 25 December 1940 against a German *Hipper*-class cruiser, HMS *Bonaventure*, a *Dido*-class cruiser, used her Type 279 as a rangefinder but found it difficult to stay on the target (she fired mainly blind ladders, and estimated range changed as much as 3300 yards in two minutes). Comments from HMS *Excellent* (March 1941) included a note that her sister ship *Naiad* was now plotting radar ranges, and that later ships would have arrangements to plot and to follow (ie, be able to compute range rates from) their radars. No such arrangements had yet been made because Type 279 was an air-search radar from which surface ranging had not been expected. Home Fleet expected that once radar had been developed further standard practice would be to use zigzags centred on radar range.[1]

Soon after the first air-search ('air-warning') radars were developed, the Admiralty Signals Establishment began work on 50cm gunnery radars, as this was the highest frequency (600 MHz) at which useful amounts of power could then be generated, ie, it would give the narrowest beams. The surface gunnery Type 284 (Type 285 was anti-aircraft) was first tested in December 1940 on board the battleship *King George V*, achieving 20,000-yard range on a cruiser.[2] Sets necessarily produced broad beams, hence were considered range-only, equivalent to optical rangefinders. Type 284 was the prime wartime main-battery radar, used in both the *Bismarck* and the *Scharnhorst* actions.

The Royal Navy was more aware than others that any electronic emissions could be intercepted. It is sometimes claimed that German surface radar gave *Bismarck* a major advantage in fighting HMS *Hood*, because the British ship had her set turned off (to avoid counter-detection). *Prince of Wales* failed to get radar ranges at all. The operator on board *King George V* initially mistook echoes of splashes for those from *Bismarck*, which were very indistinct. Shock put the ship's Type 284 radar out of action after the first thirty minutes of action.

Even so, the action was seen as a successful test of *King George V*'s combination of radar and earlier spotting concepts. The radar was integrated with the fire-control table by adding a pen showing radar range to the range plot. The target was initially found by conventional salvo patterns, using the radar as an extra precise rangefinder (the Admiralty commentator added that until the performance and reliability of radar were better understood, the size of the salvo steps used to find or regain the target could not be reduced). Once the target was found, a straddle correction was applied, each double salvo being at the radar range plus or minus that correction. Zigzags were superimposed as needed. If the target were lost (ie, if two zigzags failed to enclose it), the ship would use the normal regaining ladder. Because radar might fail at any moment, the ship kept the usual clock solution running. The commentator pointed out that straddle errors could be deduced from the difference between true range and radar pens on the plot, not at the moment of firing, but at the moment of splashing.[3]

The Admiralty was greatly impressed with the ease with which splashes could be seen; the operators needed some indicator that shot was falling, so that they could better distinguish splashes from target (*King George V* fitted a lamp). Spotting individual splashes was not enough, because ideally the mean point of impact of the salvo was wanted. Radar also raised the possibility of firing calibration salvoes to correlate gun and radar range, something impractical with optical rangefinders. Operators in *King George V* saw not only the target and the splashes, but also the shells in flight – which could distract them.

Spreads too often seemed excessive. To some extent that could be blamed on personnel fatigue and also on the flatness of the shell trajectories at the final ranges of 3000 to 4000 yards, but the Admiralty found it

The battleship HMS *Nelson*, carrying the most powerful guns in the Royal Navy, shows wartime radars in this 23 May 1944 photograph. She had received them during a refit between October 1941 and March 1942. Although she is supposed to have been fitted with a Type 284 for her main battery DCT, she appears to have an anti-aircraft Type 285 (note the 'fishbones'). A similar antenna surmounts her high-angle director, which is flanked by pompom directors carrying their own ranging radars (Type 282, recognisable by its four rather than six 'fishbones'). Also provided during this refit was a Type 273 surface-search radar (the 'lantern' visible on her mainmast). The ship had already received a Type 281 air-search set (the separate receiving and transmitting antennas of which occupy her topmasts) during a Portsmouth refit between January and August 1940. Other improvements were updated and more numerous pompom directors (three added, two on the mainmast and one at the after end of the shelter deck [01 level]), and four AA barrage directors (each with a Type 283 radar) at the after end of the superstructure abreast of and abaft the mainmast were added. The after light directors are not clearly visible here.

worrisome. The two battleships had radio communications problems, and there was apparently some interference between radar and radio, despite their use of very different frequencies.

Observation of German radar-controlled night fire against attacking British destroyers suggested that it was accurate for range but not for line. (Ships trying to shadow German capital ships always attempted to be end-on, in order to present a smaller target.) However, according to the action report, the German main-battery gunners seemed to find the range more difficult to get. In her day actions, except for that against *Hood*, *Bismarck* generally seemed to start firing well short of her target. The report speculated that this might be an attempt to spot for line without giving away the accuracy of the fire, but also that it might represent genuine failure – in which case it would be all the more important for the British to fire for initial hits using radar ranging.

Commander of the 1st Cruiser Squadron (on board HMS *Norfolk*) was:

always wondering if the enemy were using RDF [radar] to locate us, but I [correctly] have the feeling that his RDF is linked with his gun control and does not search independently. Otherwise the cruiser should have been prepared for us at 0230 on 23 [May] and on various other occasions when clearing visibility brought us in sight at ranges from nine to thirteen miles. That he does fire at unseen targets is shown by the experiences of the aircraft and destroyers when actually fired upon under these conditions.

Both wartime editions of *Progress in Gunnery* were issued with Admiralty orders warning that they often could not always distinguish splashes from targets. The

1942 edition of *Progress in Gunnery* warned that radar was not yet precise enough for blind firing. There were radar-spotting rules: if shifting to an adjacent target or reopening fire after a pause, gunners should use a 200-yard rather than a 100-yard zigzag around the corrected radar range. Plotting techniques developed pre-war surely helped convert such information to useful tactical situational awareness. This capability can be contrasted with that of enemy navies that apparently had had little or no earlier interest in tactical plotting. It also seems likely that intense interwar interest in night battle, though not directly relevant to using radar at long range, helped make the Royal Navy aware of the potential advantage radar offered.

The second generation of wartime British radars used magnetrons to generate short-wave (10cm) signals, their beams narrow enough for night fighting. Initially all production was concentrated in surface-search sets (Type 271) suitable for the small ships fighting the Battle of the Atlantic. The first was installed on board HMS *Prince of Wales* in August 1941.[4] The improved Type 273Q on board HMS *King George V* could detect a battleship at twenty-three nautical miles, and outranged Type 284. The corresponding surface gunnery set was Type 274. It could lock onto a target and track it in bearing as well as in range. Like its predecessors (and unlike the US Mk 8 and Mk 13), Type 274 pointed only at the target, hence could not spot splashes that missed for line. It was therefore supplemented with an auxiliary 'spotting radar', Type 930.[5]

The *Scharnhorst* action (26 December 1943) illustrated the strengths and weaknesses of the radars then in service. Both *Duke of York* and the cruiser *Belfast* had magnetron search radars with PPIs (Plan Position Indicators; the map-type displays now familiar), which made it much easier to control a very complex engagement involving two convoys (plus escorts), a cruiser squadron, and two destroyer divisions. The cruisers coached the battleship into position to gain radar contact with her Type 273 at 45,500 yards. The Type 284 operator picked up the target at 31,000 yards, the L18 display in the transmitting station at 30,400 yards, and the bearing tube in the transmitting station (used to keep Type 284 pointed at the target) at 25,800 yards. *Duke of York* closed, undetected by *Scharnhorst*, to within 12,000 yards before opening fire. The older radars on the destroyers sufficed to gain contact at ten to eleven nautical miles. Using radar, the destroyers gained attacking positions and hit *Scharnhorst* with fifteen torpedoes.[6] Radar was vulnerable to medium-calibre fire: an early British-cruiser salvo wrecked the forward radar array on *Scharnhorst*.

This was an unusual action, fought at night in bad weather that caused *Duke of York* to yaw badly (four or five degrees each way). The situation was simplified in that there was only one target (but there were many British ships not too far from it, so accurate plotting was vital). 'The enemy was not keen to fight and had several targets to deal with,' which meant that it was essential that, as the British had imagined after Jutland, shells be effective enough to slow the target down; *Scharnhorst* was faster than the British ships.

As before, the British were alive to the possibility that using radar would disclose their positions. Thus, according to the *Duke of York* gunnery report of the *Scharnhorst* action, *Duke of York* observed radar silence until the cruisers picked up the target. It specifically mentioned that no anti-direction-finding measures were taken while the radar (presumably Type 273) was scanning. A British attempt to jam the German radar using a Type 91 jammer aboard a cruiser failed due to mis-tuning.

Because there was only one target, the captain of *Duke of York* was not distracted by other echoes or by fear of a destroyer attack, the implication of the report being that this organisation could handle the situation but perhaps not a more complicated one. The ship's organisation for night action had been much improved after a chaotic night-encounter exercise a few days earlier, on 12 December. The ship's gunnery officer placed himself in the DCT because blind fire was considered tricky, and he wanted to be 'on the spot' if something went wrong. He wanted close coordination with the ship's captain, so both stayed on an open phone line throughout the action with the transmitting-station officer. The captain was thus aware of problems

Emerging from a refit at Philadelphia Navy Yard on 14 September 1943, the old battleship HMS *Royal Sovereign* had a Type 284 antenna atop her armoured rangefinder and the 'fishbones' of Type 285 atop the high-angle director at her masthead. The cylinder below the fire-control top was her director. The topmast carries her Type 281 air-search antenna. Not shown is the 'lantern' of her Type 271 surface-search radar, aft.

encountered at the computer in the transmitting station. He helped enormously (according to the report) by informing the transmitting station of a pending course change or avoiding such a change when a broadside was about to be fired. The report mentioned considerable 'nattering' (complaints) about the yaw; at one point the captain sarcastically asked the gunnery officer whether he thought he was creating it on purpose. The transmitting-station operator at the speed-across plot could see the bearing tube of the Type 284 radar, and he knew enemy movements before that radar picked them up.

The ship used the trainer of her DCT to track the target before opening fire, and then to follow the target based on the bearing generated by the AFCT. The trainer had thus endured at least thirty minutes' eye strain before opening fire, and he found it difficult to pick up the target despite good illumination. Later his night vision suffered because he found himself switching constantly between his binoculars and the illuminated dial showing training angle (the US Navy had something like the same problem in night battle in the Solomons).[7] With bad yaw, the trainers at the guns (and, presumably, at the DCT) must have had 'a very trying time,' according to the gunnery report.

Starshell was still important; the gunnery report pressed for a long-range type (HMS *Savage* fired her starshell with full [maximum-range] charges during the battle). In 1943 work was proceeding on shells that could be fired to 20,000 yards. Using starshell and flashless powder denied the Germans any aim point. Only when the flashless powder for secondary-battery 5.25in guns ran out did the Germans see flashes at which they could aim (the flashes also blinded those on the *Duke of York*'s bridge). Until that point the ship's captain had not even been aware that the secondary guns were firing. He had to order them to stop. During the last stages of the action, when *Scharnhorst* fired at attacking British destroyers she silhouetted herself with her own flashes (the Germans clearly did not understand the virtues of flashless powder), making her an excellent visual target and also making spotting easy when shells fell at the same time.

The battlecruiser *Seydlitz* survived turret hits at both the Dogger Bank and at Jutland; in each case her two after turrets were burned out. The Germans used brass cartridge cases rather than bagged powder, and there was considerable speculation that this practice had saved their ships. Although a magazine hit ignited all the powder, the cartridge cases burned one by one, and thus did not generate sufficient gas at any one time to destroy the ship. The battlecruiser *Derfflinger* similarly survived the loss of all her turrets at Jutland. However, when considering new cruiser guns in 1926, the British rejected cartridge cases as being no safer than their bagged cordite. This photograph was taken early in the ship's career, when she still had torpedo nets. They were eliminated because shellfire could damage their booms, leaving the nets free to snarl a ship's propellers. The net itself is the thick dark line above the line of diagonal booms. It was never to be used at sea, but rather to protect a ship when at anchor in an otherwise lightly defended port.

more of her firepower if a triple turret were put out of action. The sheer size of the triple would also cause problems (eg, it would make a larger hole in the armoured deck). On the other hand, limiting the number of turrets was attractive because it made for a shorter armoured box and better firing arcs. The weight of the next higher-calibre twin turret (15in) was about that of a triple 12in, so the Germans adopted the new gun in 1913, for the *Baden* class. They wanted a lighter weapon for the corresponding *Mackensen*-class battlecruisers: 14in.

For the Germans as for the British, Dogger Bank was a wake-up call. Ranges were far greater than had been expected, and in some cases German ships were outranged altogether. The policy of using smaller-calibre, higher-velocity guns was called into question, the Germans being very impressed by the British 13.5in guns, which did enormous damage to SMS *Seydlitz* (with one hit) and to the armoured cruiser *Blücher*, which sank. Severe damage to SMS *Seydlitz*, two of whose turrets were wiped out, caused the Germans to improve their magazine arrangements. However, they still had burned-out turrets at Jutland. The German analysis was that the British shells had been effective only when they plunged through deck armour at very long range, an eerie foretaste of the British explanation for German success at Jutland.[4] After Dogger Bank the Germans began to increase gun elevations to achieve the sorts of ranges at which they now expected to fight.[5] At Jutland, their battlecruisers opened fire at

16,200 metres (17,710 yards), and later ships were modified for longer ranges. After Jutland fleet opinion demanded a new gun more powerful than the British 15in; the proposed weapon (not yet available) was a 16.5in (420mm).[6] The naval administration in Berlin resisted this pressure, probably because it would have involved far greater expense than the existing 15in guns.

The World War I system

At the new ranges and speeds the combination of St.G. and table look-up was no longer adequate. In 1908 the Germans introduced a Dumaresq-equivalent they called an EU/SV-Anzeiger (EU for range, SV for deflection). The *anzeiger* provided initial deflection (there was no equivalent to the British bearing plot).[7] Its accuracy in turn depended on the judgement of the gunnery officer. If the target manoeuvred (or if his ship manoeuvred), the *anzeiger*, like a Dumaresq, became the only guide to current rates. The gunnery officer's judgement became more vital as the Germans switched to high-range-rate tactics from about 1911 onwards. Presumably the load imposed a considerable strain on artillery officers (no one writing later mentioned it, however). That was acceptable in the context of a quick run in before settling down at a steady 6000-metre (6560-yard) range. The run-in also explained why the Germans avoided plotting: they had to open fire instantly *as a defensive measure*. However, World War I-battlecruiser tactics in effect used only the run-on, so the strain was sustained. This strain may explain why German fire became noticeably less effective as Jutland unfolded.

The sextant was also inadequate; Zeiss began experiments with stereo rangefinding in 1900, and it first advertised the method in 1905. Stereo rangefinders were adopted in 1908 after comparative trials with a Barr & Stroud coincidence rangefinder. This choice was kept secret. Shortly before World War I the German navy decided that the smallest angle a rangefinder operator could measure was ten seconds of arc (other navies used twelve or fifteen). That set minimum range errors for the standard 3m rangefinder: sixty-five metres (71 yards) at 10,000 metres (10,930 yards) and 165 metres (180 yards) at 16,000 metres (17490 yards).

Like the original Dreyer Table, the third element of the system was designed for battle at steady range rates. It mechanically averaged ranges from all the ship's rangefinders, showing a reading from each, and calculated a range rate. The operator could throw out any reading he considered wild, and he could see the spread of the readings. Rangefinders whose data were thrown out had their lamps turned off. The device also estimated the enemy range rate on the basis of successive averaged ranges. This was how the British had calculated range rates in pre-plotting days, the main improvement being better and more frequent range readings. The Germans understood that they had to restart their rate device when turning. This kind of device had other limitations, however. Its users had no way of keeping the rate if the target was not continuously in sight. Postwar, the Germans contrasted their method, which they claimed was very quick, with what they claimed was the much slower British plotting technique.

As in a Dreyer Table, the calculated range rate was not automatically entered into the range clock; that took an order from the artillery officer. That allowed for his judgement and ensured against failure of the automatic calculator. Ranges on the clock were automatically transmitted to the guns' follow-the-pointer receivers. The experience of SMS *Derfflinger* at Jutland shows how this system could fail. Initially all her rangefinders agreed – at the wrong range. That threw her range rate off. It took the ship four minutes to find her target, after six down corrections.[8]

Fire was controlled from the conning tower, the artillery officer standing near the commanding officer of the ship so that he could receive orders. He spotted the fall of shot, assisted by other observers, and he ordered range and other corrections. No plot was maintained, but corrections were tabulated to indicate trends. A small transmitting station below the conning tower corresponded to the original British concept of a fire-control switchboard, with no computing function.

The two Badens were the most highly developed of the German World War I battleships. Baden is shown at her surrender at Scapa Flow in 1918. At this time she was missing the rangefinder atop her fire-control top, presumably removed (with all other fire-control instruments) before the ships were handed over. Presumably in response to war experience, these ships had longer-base rangefinders protruding from the forward part of their turret sides. Scuttled at Scapa Flow, Baden was raised in July 1919 and used as a target by the Royal Navy, largely to confirm the adequacy of the new generation of post-Jutland shells. These ships introduced the 15in gun to German service. Apparently they were conceived as a response to the British 13.5in gun rather than to the 15in, of which the Germans were unaware when the ships were designed. As in previous classes, the choice was a light, high-velocity shell: 750kg (1653lb) at 800 metres (2624 feet)/second. By way of contrast, the British 15in/42 fired a 1920lb shell at 2450 feet/second. The corresponding battlecruisers, which were never completed, were the Mackensen class, which would have been armed with 35cm (approximately 14in) guns, again to make it possible to build a fast battleship on about the same displacement as these ships. Although the light fast shell might seem intended for shorter ranges, these two ships had the long rangefinders (8m [26ft] base) introduced into the entire German fleet in wartime. Turret rangefinders were, unusually, near the face of each turret.

The artillery officer's periscope was target designator. At the outbreak of war ships were being fitted with follow-the-pointer receivers at the turrets; they thought of the periscope as a kind of director. However, the officer using it had no firing key; guns were fired upon a signal from a gong. It is not clear to what extent the Germans had an equivalent to the British converger, which ensured that fire from separated turrets would converge on the target at the set range. At Jutland, when visibility from the conning tower was poor, in some ships the artillery officer relied on his assistant in the foretop, who checked whether a marking atop the periscope (whose orientation in turn would point the guns) pointed to the target. This makeshift was used by *Derfflinger* when she fought *Queen Mary* at Jutland. After Jutland the Germans developed a director which could be placed in the foretop, trials being carried out on board SMS *Derfflinger*.

Typically the gunnery officer opened fire on the basis of the *anzeiger* setting, switching over to the measured range rate when it became available. Like the Royal Navy, the Imperial German navy used brackets (the British wrongly assumed it used ladders). The usual unit of range was the 'hm' (hundred metres, about 109 yards). The standard opening bracket was 8hm, about twice the standard British one.[9] The Germans thought that gave them much quicker results. As in the Royal Navy, the standard salvo was one gun per turret, although some ships fired both turret guns in quick succession.

Writing in 1930, the gunnery officer of SMS *von der Tann* recalled using an 8hm bracket against HMS *Indefatigable* at Jutland even though his rangefinder operator found all the rangefinders agreeing to within 100 metres (in effect he was taking no chances on the difference between gun range and rangefinder range). He had just corrected to 162hm (16,200 metres) range when the order came to open fire. The first salvo was correct for line but over (splashes were quite visible). He ordered 'eight down, four to the left, salvo'. These were shorts. While this salvo was in the air, he asked for the range rate. On that basis he expected his next salvo (up four) to straddle, which

The battlecruiser *Hindenburg* was considered the ultimate development of German World War I capital-ship design, although further ships of a new class (*Mackensens*) were left incomplete at the end of the war. She was completed with a tripod mast to support an enlarged fire-control top. *Hindenburg* is shown lying at Scapa Flow in November 1918. This class paralleled the *Koenig* class, the first German battleships to have all of their turrets on the centreline (and two superfiring forward). Instead of sacrificing gun calibre, they sacrificed one turret, the theory being that eight 12in/50 were at least equivalent to ten 11in/50. Broadside weight slightly increased, from 3030kg (6680lb) to 3240kg (7142lb).

it did. His target zigzagged to try to escape his fire, and he applied small corrections (1hm up or down), equivalent to the narrow ones in the later British zigzag group. The officer commented that he could estimate a change of rate by observing the target's bow, and a change in course at her bridge. Note that this account makes no mention of ladders. Nor does the published account of Commander von Hase, gunnery officer of SMS *Derfflinger*.

Although the gunnery officer could not see hits, he claimed that the ensign in the foretop could distinguish them by observing a zone of destruction spreading through the ship. Once he was clearly hitting, he ordered rapid salvoes, very much as the British would. In some cases he adjusted aim while a salvo was in the air.

Standard policy was to use both main and secondary batteries together. At Jutland, for example, a standard order called for two secondary-battery salvoes between each pair of main-battery salvoes (twenty seconds apart) on board the battlecruiser *Derfflinger*. It was apparently difficult to distinguish secondary-battery splashes, and after a time these guns were ordered to cease fire.

Like other navies, the Germans practiced concentration by pairs, using fall-of-shot indicators to distinguish two ships' splashes – as when *Seydlitz* and *Derfflinger* engaged HMS *Queen Mary* simultaneously at Jutland.[10]

By the time they began practicing gunnery in the rough open sea, the Germans had adopted design practices which gave their ships a high chance of survival, but which also made them very stiff, hence quick rollers. A typical rate of three degrees/second made continuous aim impossible. The roll was so quick that it had a noticeable effect on a gun fired at the top or bottom of a roll. The Germans' solution was a corrector, which took the time lag of firing and the roll rate into account. The contemporary Royal Navy was concerned with much the same problem, but apparently had more tractable roll rates. The corrector was first tested in 1909, and in the North Sea and North Atlantic in 1910. For the longer term the Germans bought rights to the Austrian von Petravic gyro-firing mechanism in 1909. The sight was held continuously on target by gyro. When the guns reached the correct elevation, they were fired automatically by electric contacts. This device reduced dispersion by about one third in a ship rolling twelve degrees. A few ships had been fitted with this von Petravic gyro-firing mechanism by the time of Jutland.

Probably the most original feature of the whole German system was the AC (alternating current) synchro used to transmit data. Siemens-Halske discovered that AC current made possible a self-synchronising transmitter-receiver pair that could control a dial. An AC current in a fixed rotor creates a voltage in a fixed stator (as in a transformer); in a DC (direct current) device, it takes motion to generate a voltage or a current. The amount of voltage depends on the rotor position. The transmitter was called a generator and the receiver the motor. Each, however, could function as either generator or motor. In each, the stator had three windings 120 degrees apart, like the three magnet windings in a stepping motor. The rotor was fed with the ship's AC power. Any rotor position corresponded to a particular combination of voltages (magnetic fields) in the three windings, generated by the AC current in the rotor. If the generator and motor had their rotors in the same position, then their stators felt the same voltages, and no current flowed between the two. However, if one was not in the same position as the other, the voltages would differ and current would flow. If one rotor was held in position, the other was propelled into the same position.

The Germans stripped these unique motors from their ships before turning them over to the Allies after World War I, their navy selling them back to Siemens-Halske. The French obtained many of them as reparations. In the United States General Electric independently discovered the concept, and it appeared for the first time in the US Navy in the *Maryland* class, the last of the World War I-battleship generation. Because the device was self-synchronous, General Electric called it a 'selsyn'; during World War II the US Navy changed this term to 'synchro'.

Austria-Hungary

The Austrian navy is included here because of its association with the German. Because the Austro-Hungarian empire was dissolved at the end of World War I, its navy was unable to conceal its fire-control system, and details became available.[11] The French were shocked that heavy guns were calibrated to 26,000 metres (28,433 yards) (medium weapons, as in battleship broadside batteries, to 15,000 metres/16,404 yards). They first saw AC synchros on board surrendered Austrian ships (the Austrians also used Vickers follow-the-pointer transmitters). Rangefinders were Barr & Stroud coincidence units (stereo was used only for anti-aircraft).

The US cover sheet on a report translated from an original Austrian document observes that the Austrian system was patterned on the German, but not improved in wartime, hence of little interest. The report itself suggests a very simple system, with no provision for range rates (except for one oblique comment), and also without the German automatic range averager (ranges were averaged by eye using a plot, probably of range versus time as in a Dreyer Table). There is no mention of a range clock. On the other hand, the system was similar to the German in that the gunnery officer was responsible for spots and corrections. He took target bearings and transmitted them to the rangefinder and to the turrets. There was no automatic means of correcting for parallax. Ranges were taken every thirty seconds. Salvoes consisted of one gun per turret, eg, four shots in a dreadnought with four triple turrets. This practice was explained as a way of economising on shells and also as a way of achieving high salvo rates (twenty to forty-second firing intervals). Standard practice was bracketing, the initial interval being 10 per cent of firing range or eight times the standard dispersion at that range. As in other navies, once the target had been crossed, the bracket spacing was halved, the process in this case ending when the spotter thought that 30 per cent were shorts, the rest covering the target. The writer described the system as old-fashioned, but saw wartime performance as proof that it worked.

Although the Austro-Hungarian navy disappeared with the end of the Austrian empire in 1918, it affected later events because its ships were given to both France and Italy. Impressed that the Austrians expected to hit at much greater ranges than they had imagined practical, the French were led to modernise their existing battleships. A report prepared for the US Navy by a former Austro-Hungarian officer suggested that no special fire-control equipment was involved, and it is striking that the Austro-Hungarian dreadnoughts (*Viribus Unitis* is shown, in 1912 as completed) lacked aloft fire-control positions. Ordered in 1911, they were the first battleships to use superfiring triple turrets, which enabled them to mount a more efficient battery on a limited displacement. Although allied with the Germans, the Austrians did not adopt the same methods. Thus they used Barr & Stroud coincidence rangefinders (3.65m [12ft] base in the *Viribus Unitis* class; the next [*Ersatz Monarch*] class would have had 5m [16.5ft]). They considered 3m (9.8ft) instruments effective out to 12,000 metres (about 13,100 yards). The Austrian Model 1910 (K.10, first fired May 1908) Skoda 12in/45 gun fired an unusually heavy 450kg (990lb) shell at a relatively moderate muzzle velocity of 800 metres/second (about 2625 feet/second). This was exactly opposite to German practice. However, like the Krupp guns, the heavy Austrian guns used cartridge cases and horizontal sliding-wedge breeches and hence were described, at least by the British, as Quick Loading (QL). The gun weighed 51.9 tons. In 1911 British naval intelligence reported both that a larger gun and a 12in/50 had been tested. The larger one was chosen for the next class of battleships: on the eve of war a further class of 24,500-ton ships was planned, each armed with ten 35cm/45 (about 14in/45) guns firing 710kg (1565lb) shells – again, heavier than foreign 14in shells. Two or possibly three of these guns ended up in land service during World War I.

The Austrians expected to fight in line ahead roughly parallel to the enemy fleet while steaming at similar speeds, in other words, they expected small range rates. In that case the target would apparently be static, and the very simple methods proposed, with no direct means of measuring or using range rates, might well work. Short salvo intervals might make it easier to stay on a target simply by spotting. It is possible that the references to German pre-war practice are actually to German practice before the shift to firing on the fly as the fleet ran towards decisive range.

Post-World War I conclusions

After World War I the Germans used comparative figures of hits at Jutland to demonstrate their superiority. Their official history showed 120 hits out of 3597 heavy shells fired (3.33 per cent), compared to 100 hits out of 4598 British heavy shells fired (2.17 per cent). These figures are less impressive than they appear, however, given that over a quarter of the German hits (37) were made at short range against three British cruisers – *Warrior*, *Defence*, and *Black Prince* – that were not firing back. As for their assessment of British performance, the German figures credit no heavy-calibre hits on the light cruiser *Wiesbaden* (which was sunk). They recorded eight heavy-calibre hits on the battleship *Markgraf* and nine on the battlecruiser *Derfflinger*, ascribing them to medium-calibre guns not in action against those ships. These hits would bring the British total to 117. If *Wiesbaden* and the three British cruisers are all omitted, the score becomes 117 British hits (2.54 per cent) versus 83 German (2.3 per cent), and the apparently crushing German superiority evaporates. German gunnery at Jutland seemed excellent mainly because it killed three British battlecruisers. As has been pointed out in chapter 4, the battlecruisers were sunk by hits which ought not to have been fatal, because of the way the British operated their turrets. The disasters at Jutland were own-goals, not brilliant successes for German gunnery.

British observers at Jutland noted that German fire started excellently, but got worse as the day wore on, whereas British gunnery improved over time. According to a wartime Grand Fleet summary, 'the general impression [of German fire] is that at first [it] was very rapid and accurate for range, but frequently bad for line.' Although spreads were small, no ship which survived had been hit by more than two shells in a salvo. 'After starting quickly, and establishing hitting in a very short time, the German fire fell off gradually. Whether this was due to the fact of their ships being hit caused a loss of accuracy [in] the use of a system such as the Petravic [gyro firing], or whether it was due to zigzagging to avoid being hit, or whether due to visibility conditions is not known.' The Germans had rapidly fired salvoes from one gun in each turret; at this time the Royal Navy could not have done as well, firing four- or five-gun salvoes in continuous rapid succession. A British report characterised the German fire as 'instantaneous salvoes and some form of very rapid ripple.'

German perceptions immediately after the battle, according to the Austrian naval attaché, were that British gunnery was superior, the salvoes being remarkably tight, the rangefinders superior, and the maximum ranges greater (late in the afternoon the Germans were being straddled repeatedly at 17,000 yards). The main point on which the Germans were superior was the ability to react more quickly as targets manoeuvered; the Germans blamed excessive British dependence on range clocks and an inability to change range quickly enough. The attaché attributed this to British reliance on steady sustained fire, which he contrasted to German training involving large rapid changes of range. The attaché naturally omitted the key fact, that the Germans practiced big range changes so that they could defend themselves while they ran down towards the British to fight on exactly the same steady basis at their own preferred decisive range of 6000 yards.

All equipment was stripped from ships available to the Allies postwar, even SMS *Goeben*, which was nominally in Turkish service. The Germans refused to turn fire-control details over to the Allied Control Commission responsible for disarming them. In 1921 (and again in 1926), however, the Germans tried to sell the system to the US Navy, and thus had to reveal its details. At about the same time Commander Georg von Hase, who had been gunnery officer on board SMS *Derfflinger* at Jutland, published a detailed account of his ship's fire-control system, which tallied with the confidential information the US Navy had obtained.

It seems clear in retrospect that the Germans were no more satisfied than the British. They had made relatively few hits at Jutland, and their system was even worse adapted to intermittent visibility (although it could cope better with a manoeuvering target). The 1921 decision to reveal details may have marked the

THE GERMAN NAVY

Severely limited by the Versailles Treaty, the Germans used exports to finance further fire-control development. The treaty attempted to throttle German military development by forbidding such activity, but the Germans evaded it by setting up foreign subsidiaries. They included Hazemeyer in the Netherlands, which marketed the World War I system and later ones. In 1923 Barr & Stroud, Girardelli, and Hazemeyer tendered for fire-control reconstruction of the Swedish *Sverige* class, at which time none of these companies could offer a synthetic system. Hazemeyer won the contract. The date and the ships' appearance suggest that they received the German World War I system. *Sverige* was the first to be modernised, in 1926, followed by *Drottning Victoria* in 1927 and *Gustav V* in 1930. She is shown at Stockholm in May 1934 (the photograph was taken from USS *New Orleans*). Her foretop recalls German World War I practice, but it contains a rangefinder. Presumably the large rangefinder was fitted atop her bridge because it could not fit the forward turret (the turret did have an armoured rangefinder at its rear). *Sverige* was armed with Bofors 11in guns (actually, like the Germans, 283mm): Model 1912 280mm (11in)/45s (hence comparable to the *Nassau* guns) firing 305kg (672lb) shells at 870 metres (2854 feet/second. Maximum elevation, originally twenty degrees, was reportedly increased to thirty-five degrees during reconstruction.

beginning of work on the successor synthetic rather than analytic system used during World War II.

A transitional system was installed in the pre-dreadnoughts (*Schlesien* class) which the Versailles Treaty permitted the Germans to retain. These ships had the first new-generation 5m (16.5ft) rangefinder (1926). With very little space available internally, system elements had to be concentrated in an enlarged masthead; the director and transmitting equipment were co-located. This was probably also the system installed on board the first postwar cruiser, KM *Emden*.

The Dutch connection

The Versailles Treaty prohibited German companies from selling war material, the hope being that they would shrink and that an inspiration for German militarism would disappear. Companies evaded the treaty by setting up foreign fronts. In 1921–22 Siemens & Halske of Berlin set up a fire-control subsidiary of an existing Dutch electric company, Hazemeyer (Hazemeijer) Signaalapparaten of Hengelo, which was effectively a front for them; it was managed almost entirely by Germans until the mid-1930s.[12] Hazemeyer bought its optics from a Zeiss subsidiary at Venlo. Hazemeyer apparently became the prime developer of post-World War I German naval fire control. In 1926 a director of the parent company told a US naval attaché that fire-control equipment was no longer being made in Germany, only at the Dutch plant (but in 1927 the Dutch director said that all German equipment was made in Germany). On a March 1927 visit, the US naval attaché saw several of the wartime

German fire-control devices, such as the range averager, being manufactured. About 1947 Hazemeyer was renamed Hollandse Signaalapparaten (HSA or Signaal), becoming a prominent naval radar producer; it is now Thales Naval Nederland.

By mid-1927 Hazemeyer had manufactured the fire controls for the Dutch cruisers *Java* and *Sumatra* and had a contract for six destroyer outfits (the Dutch *Evertsen* class), plus one contract for Sweden (replacement fire controls for the coastal-defence battleship *Sverige*). An outfit had been delivered to Italy (presumably it was the German destroyer system described in chapter 13, page 269). All of these systems were essentially the German World War I system, which did not incorporate any kind of computer. The company was bidding for Chilean and Argentine programmes. The company provided the anti-aircraft fire controls for the two Argentine cruisers built in Italy; the Chilean programme was presumably for the British-built *Serrano*-class destroyers (which probably used a Vickers system). Systems for the German navy were made in Germany.

The *Evertsen* system was a three-man director, described by the Royal Navy as more elaborate than that normally seen in a destroyer. The device was on a pedestal on the forebridge, using a Zeiss periscope to observe the target. The control officer, layer and trainer all used the same glass, so they were coordinated. The sight-setter faced the control officer, with his back to the target, applying the spotting corrections ordered by the control officer. The latter fired the guns. Each gun also had a local director sight.[13] There was no associated computing device.

Hazemeyer developed a close relationship with Bofors, so that ships armed with Bofors weapons during the interwar period, such as the two Finnish coastal-defence ships, almost certainly had Hazemeyer fire-control systems. By the early 1930s Hazemeyer was a prominent anti-aircraft fire-control supplier. Its stabilised, self-contained twin 40mm gun arrived in Britain on board the Dutch gunboat *Willem Van Der Zaan* when the Netherlands fell, and was copied for the Royal Navy.

The three 'K' cruisers were the first German warships to have fire-control computers, and thus were a vital transitional step to the standard World War II system. Note the masthead control position, similar to that in the 'pocket battleship' *Deutschland* and in the Finnish coast-defence ships. The rangefinders lack the elaborate multiple functions of the later units. The head of a periscopic director is visible abaft the masthead rangefinder. The triple turrets, which could elevate to forty degrees, fired a relatively light shell (45.5kg/100lb) at high velocity (960 metres [3150 feet]/second). Maximum firing rate was eight rounds/minute per gun. This 5.9in/57.5 C/25 gun was only one of a range of German 5.9in weapons used at sea during World War II, others being the standard capital-ship secondary gun (5.9in/52.4 SKC/28, firing a slightly lighter shell) and a destroyer gun (5.9in/45.7 Tbts KC/36, firing a 40kg [88lb] shell). The German World War I 5.9in/42.4 SKL/45 fired a 45.3kg (99lb) shell at 835 metres (2740 feet)/second. It armed light cruisers and formed the battleship secondary battery – which was to have been used against battleships.

Photographed before World War II, the 'pocket battleship' *Admiral Scheer* displays her rangefinders and also her periscopic directors. The head of one of them is visible on the fire-control platform just forward of the masthead rangefinder. These were the standard German directors of the World War II period, descended directly from the World War I type. Each contained three periscopes: for the gunnery officer, for the trainer, and for the layer. Weight was reduced by separating the transmitting unit (to the guns) from the director proper, so several could be placed together. This separation came in the 1930 version of the system. Given this separation, any director could be connected to any gun via a fire-control switchboard; the directors were said to be 'neutralised' for gun calibre. Unlike contemporary directors in other navies, the German directors made no provision for cross-levelling, which was done entirely by the ship's stable element.

The World War II system

In 1926 Hazemeyer's sister company Nedinsco (for optics) hired Barr & Stroud's fire-control consultant, who had been in charge of their system development, the retired Dutch Admiral Mouton. Barr & Stroud asked him not to take this post (to no avail), on the grounds that he would be developing fire-control systems for the Germans who really owned the company. The change in Hazemeyer systems was dramatic. As of 1927 the company was not producing anything like a range-keeper; a US attaché report on the system on board the Dutch cruiser *Sumatra* said specifically that no tracking instruments were used.

About 1926 the Germans adopted graphic plotting. This date is consistent with the beginning of development of the analogue gunnery computer (*schusswertrechner*) which the Germans first placed in service in 1930 on board the 'K'-class light cruisers (C/30 version).[14] It also equipped the light cruiser *Leipzig* and the 'pocket battleship' *Deutschland*. In these ships the range and bearing plots were separate from the computer proper, which was in the form of a table like those of British systems. This type of computer apparently also equipped the Dutch cruiser *De Ruyter*.[15] *Nuremberg* and the two later 'pocket battleships' had C/32.

The *Scharnhorst*s had the C/35 version of the computer, with all new auxiliaries. It was probably the first to have the two plots integrated into its face. In arrangement the new computers resembled the Italian computers derived from earlier Barr & Stroud practice. It may be relevant that in 1931 the Italians provided the Germans with a set of their fire-control equipment. However, without a detailed description of the mechanism of the German computer it is impossible to say whether it was related to Barr & Stroud's. The final German computer was C/38: the battleship (S) version for the *Bismarck*s and the cruiser (K) version for the *Hipper*s, including their sole survivor *Prinz Eugen*.[16]

By 1927 Hazemeyer was producing a device to control gun elevation electrically, although it could not be shown to a visiting US attaché. He was permitted to see the electric range clock, an essential element of the system, which the Germans had had during World War I. The elevation from a directorscope was superimposed on the angle given by the clock corresponding to the predicted range, and the combination transmitted to the guns. The attaché was sceptical that remote control could overcome the inertia that would cause a gun to overshoot the desired elevation. However, it seems clear that such remote control became a standard feature of German World War II practice.

Photographs of a typical World War II German range-keeper (ie, computer) console show separate own and enemy dials. There were separate range- and bearing-rate plots (apparently conventional rather than straight-line), each showing both input data and generated data, for comparison. An operator entered each rate into the computer by moving a hand wheel, which moved a bar to parallel the apparent slope of the plot. Given range and bearing rates, the computer moved wires above the enemy course and speed dial according to range rate and computed rate-across. Using these wires, an operator could set the corresponding enemy course and speed by what amounted to a cross-cut, as in other contemporary systems. This arrangement seems to have been taken from Italian practice; the Italians presented the Germans with a

At Montevideo after the Battle of the River Plate, *Graf Spee* displayed a fire-control radar antenna on her forward rangefinder mount. The appearance of this antenna was apparently the first indication the Royal Navy had that the Germans also had radar. The set contributed little to the battle because it was soon disabled by the ship's vibration. A Seetakt radar had first been installed on board *Graf Spee* for trials in 1936. Large ships could be detected at ten nautical miles (20,000 yards), cruisers at six. In common with other contemporary radars, it could not use the same elements to transmit and to receive, so the upper part was used for transmission and the lower for reception. The reception part was divided in two for more accurate bearing measurement; accuracy was given as 0.2 degrees. By 1939 Seetakt had been replaced by the standard production FuMO 22, which in *Graf Spee* had a 1.8 x 0.8m antenna. Peak power was increased from seven to eight kW. Other 'pocket battleships' had larger antennas, eg 6 x 2m in *Lutzow* (ex *Deutschland*) in 1939 and 4 x 2m in *Admiral Scheer* in 1940. These antennas were all split into transmitting and receiving elements. They operated on about 82cm wavelength, compared to about 50cm for contemporary British sets such as Type 274. Only in 1944–45 did the Germans manage to increase power noticeably.

complete fire-control system in 1931. The computer generated range and target bearing; range was displayed on a scale between enemy and own-ship dials. An operator entered spots using a range wheel. Another large dial atop the range dial probably showed compass heading. Smaller windows and dials indicated data such as wind force, direction and latitude, which were automatically fed into the system.

As in the World War I system, the output of the computer was not automatically entered on the sights. Instead, it had to be passed from the position-keeping computer into a separate ballistic computer, using follow-ups. Presumably this separation made it possible

The German battleship *Gniesenau* is shown at or just before the outbreak of war, with what appears to be an experimental radar antenna atop her foretop director. The antenna (apparently installed upon completion) and the funnel cap suggest that the photograph was taken just after the major 1939 refit in which the ship was fitted with a sharply raked 'Atlantic' bow to reduce her serious wetness forward. At the foretop is a 10.5m (34.5ft) rangefinder (similar units were in the turrets and in the after main-battery control position); the lower rangefinder is the 6m (19.6ft) unit standard in cruisers. As in later German capital ships, the upper mast level under the rangefinder was the foretop gunnery-control position. Below it is a searchlight platform (at this level inside the mast was an air-defence control position). Below was a communications centre. The prominent platform below the searchlight platform was the admiral's bridge. The upper level of the conning tower, surmounted by the forward rangefinder, was the forward gunnery-control position. It was surrounded by the ship's bridge. These ships were armed with 11in rather than 15in guns in hopes of avoiding immediate political conflict with the British, who had just agreed to their construction by signing the 1935 naval treaty with Germany (the naval staff considered 33, 35, and 38cm guns [ie, 13, 14, and 15in]), but it was understood that only the 11in would be approved. Once war began, plans were drawn to replace each triple 11in turret with a twin 15in, and *Gniesnau* was in Gdansk for this refit when she was irreparably damaged by bombing. The black blobs surrounding the foretop rangefinder are the heads of periscope directors. This class introduced the spherically shielded, triaxially stabilised anti-aircraft director (SL6, Model 1933; the previous SL1 type was triaxially stabilised, but not shielded this way). One is visible under the bridge wings. It has a 4m (13ft) rangefinder. The guns were designated 28cm/L54.5 Model (C) 1934. The turret incorporated a delay coil for the centre gun. The AP shell weighed 330kg (727lb), compared to 300kg (661lb) in the 'pocket battleships' with Model 1928 guns. Muzzle velocity was 890 metres (2920 feet)/second, compared to 910 metres (2986 feet)/second for the earlier gun firing the lighter shell. Maximum elevation was forty degrees. The Germans claimed that at seventy degrees target angle (presumably the standard for comparison) the AP shells from these ships could penetrate 348mm (13.7in) of armour at 10,000 metres (10,930 yards) or 280mm (11in) at 15,000 metres (16,400 yards) or 225mm (8.8in) at 20,000 metres (21,870 yards). The ships were more heavily armoured than the *Bismarck*s (presumably because they had much lighter main batteries, even in proportion to their smaller displacements), with 350mm (13.7in) belt armour which, under the usual rule of making belt armour equal to gun calibre, would have been appropriate for the treaty battleships (armed with 14in guns) permissible when they were designed.

to reject a position-keeper solution. Ballistics for particular guns were embodied in three-dimensional cams in the ballistic computer; the position-keeper was standard for all heavy ships. The separation between position-keeper and ballistic computer led an official postwar British writer to say that the position-keeper was not actually part of the fire-control chain. The computer system (including ballistics) was much simpler than an AFCT, requiring only about five operators.

As in British practice, the gunnery officer was stationed at the director, rather than in a conning tower or at the computer. He had the old EU/SV-Anzeiger and a range clock, as well as own-ship instruments and a gun-ready indicator, and he had the firing pistol. As in World War I, gyro data were inserted into the guns rather than via a director; the director was much like the very simple type used in World War I. Any director could control any calibre of gun.

US and British officers who inspected German ships at the end of the war were impressed mostly by the elaborate stabilisation of German directors and weapons and (at least in the US case) by German use of magnetic rather than electronic amplifiers.[17]

The heavy cruiser *Prinz Eugen*, examined at the end of the war, had an unusually large fire-control space, with two main-battery plotting rooms, two anti-aircraft

Late in World War II, the heavy cruiser *Prinz Eugen* displays her mattress-type fire-control radar and her two main rangefinders. They were not British-style DCTs. Barely visible forward of the conning-tower top rangefinder is the stubby head of a periscope-type director; another is silhouetted at the fore end of the fire-control top surmounted by the other director. The top of the mast carries what looks like a crow's-nest, but is actually the enclosed antenna of the only microwave surface-search radar produced by the wartime German navy, 'Berlin' (FuMO 81). Its key element, its magnetron, was derived from one recovered in 1943 from a crashed British bomber carrying a microwave ground-mapping (navigation) set, H2S. The Germans reportedly deliberately made the silhouettes of their *Hipper*-class cruisers and *Bismarck*-class battleships similar, to confuse enemy observers (but the similarity might also be attributed to a single design office preparing several designs in quick succession). It has been suggested that it took HMS *Hood* time to register on *Bismarck* because initially her spotters misidentified *Prinz Eugen* as *Bismarck*. They would have set the wrong length on their inclinometer, generating the wrong solution for target course. That might explain why the initial salvoes were badly off in line (bearing) as well as in range. These heavy cruisers fired relatively heavy shells (122kg [269lb]) at high velocity (925 metres [3054 feet]/second). The forward director carries the standard German FuMO 26 main-battery fire-control radar. Despite its appearance, this was not an air-search radar. The Germans had no equivalents to the microwave gunnery sets the US Navy and the Royal Navy introduced from 1943 on (the US Mk 8 and Mk 13 and the British Type 274). Instead, the Germans continued to operate at 80cm wavelength, somewhat longer (hence less effective at low elevations) than the 50cm of the British Type 284 and its relatives and the 40cm of the US Mk 3. It had a new 6.6 x 3.2m antenna, and by 1945 had been upgraded to a peak power of sixty kW (four microsec pulses) to achieve range accuracy of five yards and bearing accuracy of 0.25 degrees. Some sets (probably not this one) were further upgraded to FuMO 34, with a peak power of 125 kW, giving them a range of forty to fifty kilometres.

plotting rooms, two master-stable element rooms, four main-battery directors (compared to two in US heavy cruisers), four anti-aircraft directors, a torpedo computing room and torpedo directors. Mechanisation was described as unusually extensive, eg, two large panels of about thirty cubic feet combined to indicate whether a selected sequence of torpedo fire was permissable. Compared to US practice, the Germans used more optics (eg, more rangefinders), and they could switch between functions (eg, main and anti-aircraft batteries, torpedoes, even navigational). The German devices were considered more complex than US optics. Under development at the end of the war was a means of measuring the muzzle velocity of each shot, using a cylindrical surface on each 8in gun.[18]

At the fore masthead was the main-battery director station, surmounted by a rangefinder station, and then by a radar room.[19] The director station contained two directors (side by side) and a total of eight crew, sufficient to operate one director at a time.[20] Each pedestal unit was served by a tube carrying three periscopes (for the three operators) through the roof of the armoured space. The unit was operated by a pointer and a trainer and supervised by the gunnery officer (a commander), acting as spotting officer. A unique feature was the mouthpiece firing key, blown into by the pointer to fire.

In a Norwegian fjord in 1942, Tirpitz displays her rangefinders and her gunnery radars. The foretop rangefinder had a 10.5m (34.5ft) base, but the unit atop the conning tower had a 7m (23ft) base, as in a heavy cruiser. The rangefinder atop the tower foremast is topped by a radar office serving the new mattress antenna on its face. Note the other antenna just below it. The original antennas on the two rangefinders were for FuMO 23 (4 x 2m). This was the original production surface fire-control set, initially designated FMG 39G(gP), indicating the year of introduction. The additional antenna on the new radar office was FuMO 27 (also 4 x 2 m). It was a 1940-series radar for surface fire control. Ultimately the two separate foretop antennas were replaced by a single 6.6 x 3.2m mattress for FuMO 26. It was initially credited with a range of twenty to twenty-five kilometres. The rangefinder atop the conning tower has no additional office because it would block the view from the bridge. Forward of the rangefinder the heads of three periscope directors are visible. The cupola atop the radar office is for air-defence control. The vertical pipes are smoke generators. The ship's main searchlight, atop her mast, is turned away from the camera.

The big sphere was a triaxially-stabilised C/37 medium-calibre anti-aircraft director. In 1944 it was fitted with a 3m (9.8ft) diameter dish antenna for the standard Wurzburg D anti-aircraft fire control radar (FuMO 213). Other German capital ships had similar radars. Thus in November 1939 Scharnhorst had a 6 x 2m mattress on her foretop for FuMO 22 (originally FMG 39G(gO)). In the summer of 1941 she had 4 x 2m mattresses both on her foretop and on her after main-battery director for FuMO 27. She had a radar office atop her foretop rangefinder from the beginning (Bismarck had no such addition). All of the 1939 radars had similar performance (FuMO 21 was for destroyers and cruisers), with a peak power of eight kW. FuMO 22 was credited with accurate bearings within five degrees and with the ability to detect a battleship at thirteen nautical miles (about 23,800 metres). The 1940 series had similar power output, hence presumably similar effective range. These ships fired an 800kg (1763lb) shell at 820 metres (2690 feet)/second, continuing the standard German practice of using lightweight shells.
PHOTOGRAPH COURTESY OF JOHN ASMUSSEN.

Optics were stabilised automatically in bearing and level by the master stable element, with twin back-up gyros.

The 7m (23ft) rangefinder above the director was stabilised by twin gyros in the event the master unit failed.[21] It had four personnel: pointer and trainer, rangefinder operator and control officer (who could be a petty officer).

Range was continuously received at the computer and displayed on six separate graphic time charts. Range rate was derived from the slope of the range-time graph, then inserted into the computer. The computer calculated the sight angle (ie, range) and deflection, transmitting them to the gun order units

The standard German C/38 computer is shown in the forward plotting room of the battleship *Bismarck*. It shows the own-(left) and enemy-ship dials typical of synthetic fire-control systems, with the line of sight (in effect) connecting them. The dial between the ships indicated the firing range and bearing. Cross-wires on the target dial were used for cross-cuts. The two glassed-in ports cover plots of target range and bearing. There were three large wheels or cranks under the computer: one on the left to enter range corrections, and two to enter range and bearing rates observed on the plots into the computer. Presumably the wheel at centre was intended to enter target course (target speed seems to have been dialled in using a dial to the right of the wheel). Recently it has been pointed out that *Bismarck* had not been completely worked-up before her sortie. If that is true, her ability to hit HMS *Hood* (and to come close to hitting HMS *Prince of Wales*) in her first battle may be tribute to the automation built into her fire-control system, just as the success by HMS *Prince of Wales* showed what British automation could do. In that case the ship's failure to continue to hit after her initial successes may point to inherent problems with the system and, in the final battle, to limits of stabilisation (she was rolling as she continuously turned).
PHOTOGRAPH COURTESY OF JOHN ASMUSSEN.

(Rw Hw-Geber), which added tilt corrections based on signals from the master stable element. The plot housed a dummy director and recorder for training. Guns followed ordered elevation automatically but were hand-trained as ordered. The plotting room required eight personnel, two of them handling bearing and range data.

As in other navies, there was a corresponding destroyer system, in this case a director with an integrated computer.

An account by the gunnery officer of KM *Bismarck* shows how the system was used.[22] He could fire either a ladder (three salvoes spaced 400 metres (437 yards) apart) or ranging shots ('test shots'). The ship could fire either half-salvoes (four guns) or full salvoes (all eight guns). *Bismarck* generally used ladders, and generally straddled on the first ladder. HMS *Hood* was hit at 16,000-metre (17,497-yard) range. She fired a total of ninety-three rounds in six minutes, but several salvoes were probably in the air when *Hood* exploded.

German wartime radars were primitive by Allied standards, and they never attained sufficient precision in bearing to make blind-fire possible. Thus lessons drawn by the British from the *Bismarck* action included the comment that when trying to fire blind the ship made errors in line rather than in range, so that shadowing ships should try to be end-on.[23]

Spreads seemed large, eg, when *Bismarck* fired at the shadowing cruiser HMS *Suffolk* on 24 May 1941 (her shells fell short). Spreads grew noticeably after the first few minutes. During her first action, *Bismarck* straddled the two British ships early, but a British commentator observed that after HMS *Hood* was sunk *Prince of Wales* managed to maintain fire with remarkably little damage except for an unlucky hit on her bridge. To two radar salvoes from *Prince of Wales* on 25 May, *Bismarck* fired first a single gun salvo, then a four-gun salvo, both a long way short, at about 16,000 yards. *Prince of Wales* fired some time before the reply. During the final daylight action, *Bismarck*'s first salvo or salvoes were usually well short. The British wondered whether this was deliberate, in order to find the line without warning the target to change course. The ship's accuracy fell noticeably as soon as she was straddled, but by that time she had been badly damaged. Rear Admiral 1st Cruiser Squadron commented that, except for the opening action, 'enemy's shooting was distinctly poor.'

CHAPTER 9
The US Navy

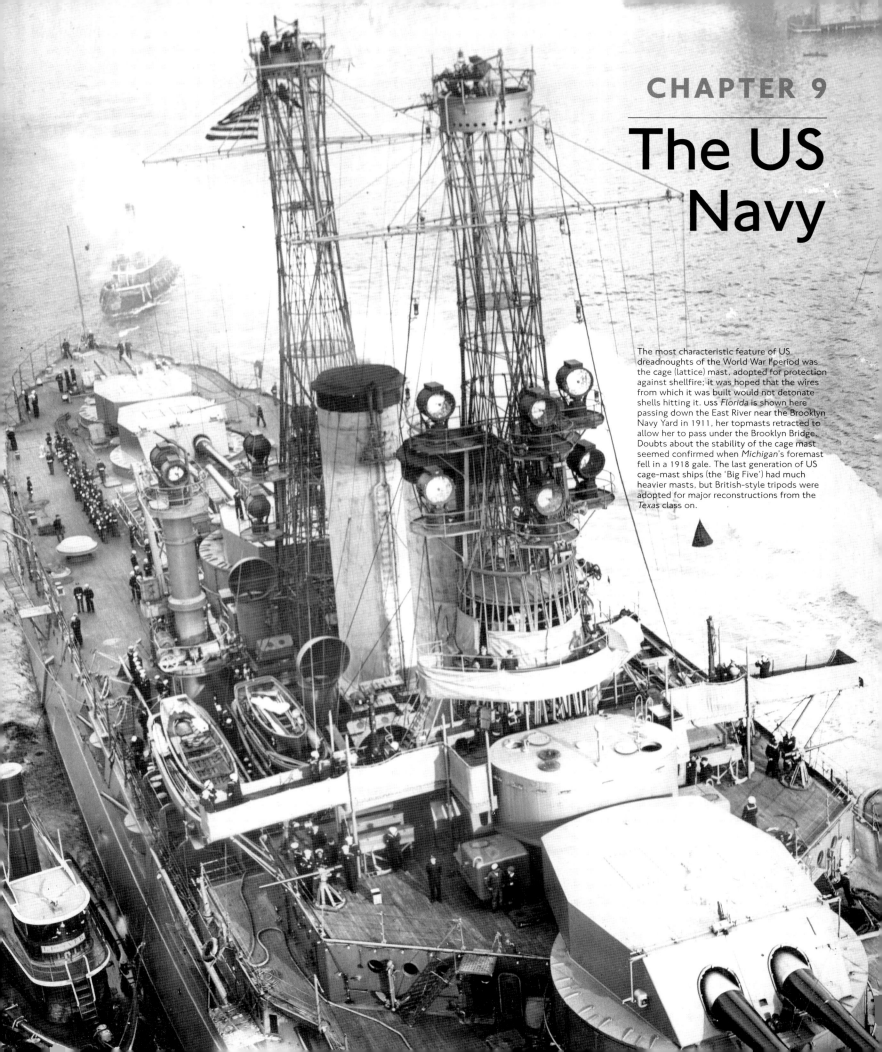

The most characteristic feature of US dreadnoughts of the World War I period was the cage (lattice) mast, adopted for protection against shellfire; it was hoped that the wires from which it was built would not detonate shells hitting it. USS *Florida* is shown here passing down the East River near the Brooklyn Navy Yard in 1911, her topmasts retracted to allow her to pass under the Brooklyn Bridge. Doubts about the stability of the cage mast seemed confirmed when *Michigan*'s foremast fell in a 1918 gale. The last generation of US cage-mast ships (the 'Big Five') had much heavier masts, but British-style tripods were adopted for major reconstructions from the *Texas* class on.

MODERN US NAVAL GUNNERY began when Lieutenant William S Sims met Percy Scott (see chapter 1) on the China Station, and saw what his revolutionary thinking could achieve. The US Navy had recently emerged victorious from the Spanish-American War, but within a few years discovered that its shooting had been quite poor, roughly comparable to that of the Chinese in the Sino-Japanese War a few years earlier. Sims dramatically demonstrated Scott's continuous aim on the China Station in 1901–2, and the technique was adopted officially in 1902 with the personal support of President Theodore Roosevelt.[1] As in the Royal Navy, continuous aim for medium-calibre guns led to attempts to provide heavy guns with controls sensitive enough for similar operation; these were tested in 1907 and installed within a few years. Like Scott, Sims was made Inspector of Target Practice, in effect the navy's chief gunner, despite his junior rank. Sims introduced Scott's systematic competitive target practice. As in the Royal Navy, it created a rush of new fire-control concepts and devices. For example, the US Navy's plotting board was an individual officer's pet idea. Sims' personal connection with Scott gave him some access to the new British fire-control concepts and experiments.

Sims was apparently responsible for longer-range experiments, beginning in 1904 with the battleship *Alabama*. They emphasised spotting, which was then Scott's solution to long-range firing. The US Navy distinguished between horizontal and vertical spotting. Horizontal spotting was direct estimation of the distance by which a shell fell short (distances over were estimated from the apparent height by which the shell missed vertically). A vertical spotter estimated the apparent distance of the slick left by a splash below the target waterline (if the slick of an over extended beyond the target, he could estimate its height), comparing it with a known vertical distance on the target. Given his known height above the sea and the approximate range, he could find the distance between splash and target. The Royal Navy discarded this idea on the ground that a shell hitting the top of a wave would seem much closer to the target; the US Navy, operating in calmer water, may not have been as concerned. In 1905 fire-control parties produced diagrams showing the range errors associated with various apparent distances below the target waterline. Instructions showing how to produce similar diagrams were a feature of all later US gunnery handbooks.

When vertical spotting was possible, bracketing (which the trials board called a fork) was not needed; a good spotter could place a well-calibrated gun (whose pointer compensated for the ship's roll) on target within three or four shots.[2] Forks were needed at longer ranges. The tests made no attempt to deal with the problem of a moving target (range-keeping), but the trials board observed that it would be difficult to set up a fork if the range were varying.[3] Sims later advocated vertical spotting as a useful way to measure the difference between rangefinder and gun range. In a 1908 essay, he argued for relying on spotting. For example, firing at a moving target, several guns could be set to fire at different ranges around the original range, to take account of different estimates of the range rate.[4]

As in the Royal Navy, the most important conclusion in 1904 was that ships needed centralised fire control from positions aloft equipped with rangefinders, in direct communication with the guns. The trial board's ideas were elaborated and extended by a special fire-control board (on which Sims served) convened in November 1905 by the Secretary of the Navy.[5] Fire control required salvo fire and a calculated range rate. Two new instruments would be used: a range projector (like a Dumaresq), indicating the number of seconds for the range to change by fifty yards; and a range clock ('range-keeper').[6] Once the ship was taking ranges, plots would give enemy course and speed 'with considerable accuracy'. The clock and range projector were calibrated from 2000 to 12,000 yards, with speeds of zero to forty knots. This did not mean a battle range as great as 12,000 yards, but rather that the instrument might be set and used as the ship approached gun range. As in the emerging British system, the guns would receive their range data from a transmitting station, in this case an annexe to the 'central' containing

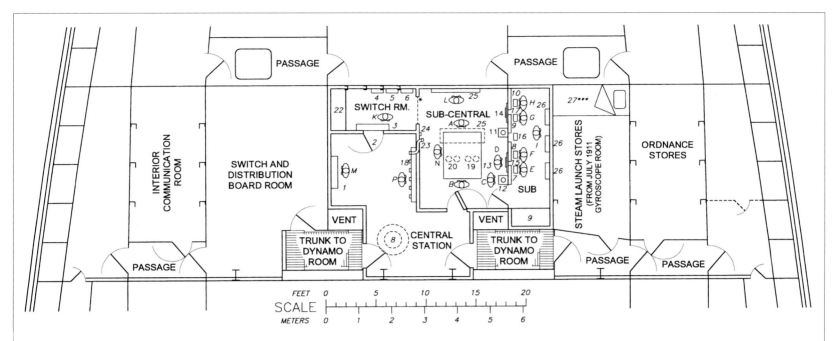

The battleship *Delaware* shows the US Navy's plotting-room arrangements before it adopted the Ford Range-keeper, an analogue computer. The space is built around a large plotting table with two plotters (A and B) and a supervising officer (N); data flowed into it from the voice tubes overhead (19 and 20). When the original of this drawing was made, the spotting tops carried the ship's rangefinders, so the tubes provided both ranges and spotting corrections. Given rates derived from the plot, two range clocks (11 and 12) could be set. This arrangement, designed as the range projector, the US equivalent of the Dumaresq, was being discarded. This drawing originally illustrated 'Questions on the Effectiveness of US Navy Battleship Gunnery: Notes on the Origins of US Navy Gun Fire-Control System Range Keepers,' Pt. 3, by C C Wright, *Warship International*, Vol 42, No 1.
(W J JURENS)

1 Ship's Service Telephone Switchboard
2 Access To Rear of FC Switchboard
3 Fire Control Switchboard
4 Five-Inch Salvo Switch Panel
5 Twelve-Inch Salvo Switch Panel (Bells)
6 Twelve-Inch Salvo Switch Panel (Buzzers)
7 Deflection Repeaters Subtarget
8 Conning Tower Tube (Over)
9 Telephone Locker
10 Range Repeaters Subtarget
11 Range Clock
12 Range Clock
13 Range, Deflection, and Angle of Train Dials
14 Range, Deflection, and Angle of Train Dials
15 Range and Deflection Transmitters Subtarget
16 Master Control Switch
17 Range And Deflection Transmitters Subtarget
18 Voice Tube Panel, Ship Control
19 Voice Tubes to After Spotters' Top, 3-Inch
20 Voice Tubes to Forward Spotters' Top, 3-Inch
21 Blackboard and Rate of Change of Range Finder
22 Work Bench and Tool Locker
23 Voice Tube to Central
24 Voice Tube to Motor Generator Room
25 Four Light Reflector
26 Connection Boxes
27 Berth (probably added to allow continuous civilian supervision of gyroscope)

Notes
* Original sketch identifies an archway' between the switch room and sub-central.
** Original document is indistinct, but probably 'N' rather than 'H'. If 'N' then perhaps an officer supervisor of plotting.
*** Probably added to allow continuous civilian/naval supervision of gyroscope.

A Plotter
B Plotter
C Clock Operator
D Dial Operator
E Switchboard Operator
F Deflection Transmitter Operator
G Turret Telephone and Voice Tube Operator
H Electrician
I Officer in Charge of Sub
J Unused
K Fire Control Switchboard Operator
L Officer in Charge of Sub-Central (?)
M Ship's Service Telephone Switchboard Operator
N Plotting Officer**
O Unused
P Voice Tube Operator

protected ship controls for use in battle (as commanded from the conning tower). The US Navy adopted 'visuals', counters which changed 100 yards for each turn of a handle (the standard increment was fifty yards).

The 1905 board proposed cage (lattice) masts, which it hoped would resist battle damage. The new fire-control arrangement was first installed on board the pre-dreadnought battleship *Virginia*, chosen because it was still under construction. Cage masts were the most unique feature of US capital ships until World War II.[7] The first permanent fire-control systems were installed in 1908. They failed during the 1909 gunnery practice, when the target as well as the firing ship were moving.

Some ships discarded both clock and projector in favour of plotting, the object being to measure rather than guess enemy course and speed (Pollen was making much the same case in England). Plotting had nearly replaced the earlier clock and projector by 1911. The US Navy used manual plotting boards: Mk II ('time-range-spot') for range versus time (ranges from 4000 to 30,000 yards, as it was used in the early 1920s), as in a Dreyer Table; and Mk III (later replaced by Mk IV) for true-course plots (called target tracking, not plotting, in the US Navy). Mk IV incorporated a pair of universal drafting machines to measure target course.[8] Both manual plotting boards were retained after a computer was introduced, not least because they offered backup in case of computer failure. The emphasis on plotting probably explains why, in 1910, the chief of the Bureau of Navigation stated that gyro-compasses were needed to exploit advances in gunnery.[9] True-course plotting seems to have been much more successful than in the Royal Navy, perhaps because US exercises were held in calmer water in which yawing was not a major factor, and in which compass performance was thus not nearly so crucial. By 1912 the desired battle range was 10,000 yards, which was about what the Royal Navy was using. Fire would open at 12,000 yards or more. The annexe was renamed the plotting room; it became the core of the US system. In contrast to the Royal Navy, control was withdrawn from the aloft platforms because they were considered too vulnerable. They were soon limited to a rangefinder and visual and telephone connections with the annexe.

Plotting made rangefinder precision a limiting factor in fire control. The US Navy initially bought Barr & Stroud rangefinders, but it wanted a domestic supplier. In 1908 Bausch & Lomb bought licences from Zeiss (for coincidence rangefinders like Barr & Stroud's).[10] In 1910, when each battleship had a 9ft Barr & Stroud at each masthead, the navy bought thirty 3m (about 10ft) Bausch & Lomb Invar rangefinders (using inverted images for comparison, rather than vertical cuts) to replace them. In 1911 a 15ft Barr & Stroud rangefinder was ordered for the fleet flagship. Then the two latest battleships, *New York* and *Texas*, were each equipped with 20 to 22ft Bausch & Lomb turret rangefinders (in their superfiring turrets). Later battleships received 26.5ft rangefinders (*Pennsylvania* and *Idaho* classes) and then 30 to 33ft instruments (*California* and *Maryland* classes). Postwar the turret rangefinders were called 'battle rangefinders' because they were armoured. Ships without rangefinders in their turrets had them mounted on top, but turret tops were needed for anti-aircraft guns and then for aircraft flying-off platforms. By 1925 long-range control depended entirely on the big turret units. Ships also had two shorter-based rangefinders in exposed positions, fore and aft (often one atop the bridge, one atop X turret), for secondary-battery and torpedo control, for ship control (navigation), and for flag use (to aid in plotting the movements of accompanying ships). Unlike the Royal Navy, until the 1930s (and the Mk 8 range-keeper) the US Navy made no attempt to register rangefinder readings automatically, relying instead on telephones. As in the Royal Navy, human plotters averaged their results by eye.

The US Navy associated long-base rangefinders with increasing battle ranges. When its battleships joined the Grand Fleet in 1917, their officers thought that by using only short-base instruments the British were try-

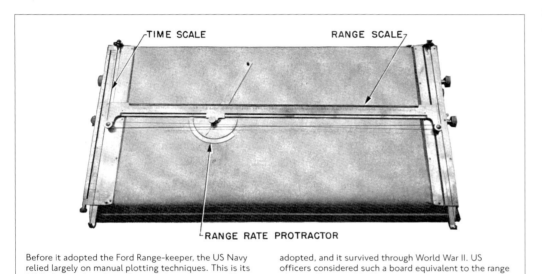

Before it adopted the Ford Range-keeper, the US Navy relied largely on manual plotting techniques. This is its Mk II plotting board, used to calculate range rate. This board was retained as a back-up after the Ford had been adopted, and it survived through World War II. US officers considered such a board equivalent to the range plot of a Dreyer table. This illustration is from the 1950 edition of the US Navy's gunnery manual.

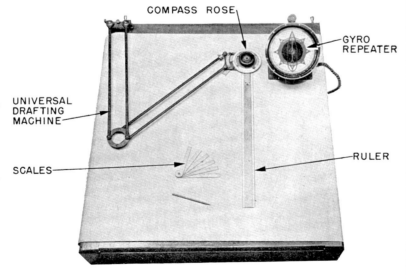

For the US Navy, plotting meant range plotting. Tracking meant what the British called true-course plotting. This illustration of the US Navy's standard tracking board is from the 1950 edition of its gunnery manual. US experience with such trackers may have been happier than the Royal Navy's because they were adopted at the same time as gyro compasses (not the gyro repeater on the board), hence were less liable to error due to yawing.

ing to 'make bricks without straw'. Post-World War I, the US Navy tested stereo rangefinders in view of German war experience. The Germans strenuously denied the British claim that stereo operators lost their ability due to fatigue or the stress of battle.[11] Trials with imported Zeiss instruments in the early 1920s seem to have been inconclusive, as the type was not adopted at this time. Stereo ranging was adopted in the 1930s.

In 1912 the Bureau of Ordnance (BuOrd) learned that the British had achieved remarkable results by director firing. Without details of the British system, it began work on its own. The first directorscope was a modified turret sight using Cory visuals (which had numerical indicators rather than follow-the-pointer) transmission. Tested on board USS *Delaware* in September 1913, it produced exceptional results, though not as good as those on HMS *Thunderer*. A similar device installed by the ship's crew of USS *Michigan* (using an ordinary turret periscope) was completed in December 1914, achieving phenomenal results in the spring 1915 target practice. This directorscope was clamped in position, the operator pressing the firing key as the ship's roll brought his cross-hairs onto the target. Directors were installed on the battleships *New York*, *Florida*, *Utah*, *Arkansas* and *South Carolina* in 1915. In smooth-water tests, USS *Texas* showed that it was enough to fire all of the guns with a master key; the director aloft was needed to handle rough weather. In 1916 the director ships fired the first US long-range practice, at 20,000 to 18,000 yards. Director control was authorised for all battleships from the *Virginia* class onward and for all armoured cruisers.

That year the Royal Navy provided more detailed director information to the US Navy, and the lessons of Jutland were analysed.[12] It was now clear that a director had to be target designator. The design of a new system was completed in December 1916, tested in January, and installed on board USS *Arkansas* in March 1917 (it was modified to final form in July). Completion was rushed for the other dreadnought battleships, with plans for installation in the later pre-dreadnoughts (*Virginia* and *Connecticut* classes) in 1918. Just before the United States entered World War I in April 1917 a simple system was designed and authorised for the earlier pre-dreadnoughts. The Royal Navy provided considerably more information once the United States entered the war.

The US directorscopes were much less elaborate than the British ones devised by Scott. They were not dummy guns, so range data did not pass through them to the guns. Their single telescopes were movable in train and in elevation. The single operator pointed his telescope at the target and 'centred the roll' (located the point around which his scope moved about equal distances above and below as the ship rolled). The director correction was read off a scale and passed to the plot (later it was transmitted automatically). The operator pressed a 'stand-by' buzzer at the top of the roll and the master firing key as his cross-hairs passed over the target.[13] Battleships generally had a director in their tops and another (using a periscope) in their superfiring turret roofs.[14] The *California* class introduced a periscope director (target-bearing transmitter) in the roof of the conning tower.

A 1917 conference with the British brought out the need for director control for secondary batteries. Vickers, who produced the British secondary-battery control system, could not provide material in time, so plans were obtained in October 1917 and the equipment produced by New York Navy Yard and by some private firms (the first complete installation was on USS *Arkansas* in July 1918, before she joined the Grand Fleet). This Mk 7 director was installed on board

This Mk II periscope director (directorscope) equipped US battleships from the last pre-dreadnoughts (*Louisiana* class) onwards through the *New Mexico* class. This photograph is from the 1918 BuOrd director manual. The periscope measured and transmitted the director correction, ie, the measure of the ship's roll angle, by tilting in the vertical plane. The directorscope also carried a warning buzzer key and a firing key. There was no correction for the height of the device. Directorscopes were installed in the tops, in the fire-control tower (part of the conning tower), and in the high turrets. Relative target bearing was measured by a separate target-bearing transmitter. The 1919 Fire-Control Board wanted both functions combined (and also combined with the spotting glass) and brought aloft to the tops, periscope directors remaining in the fire-control tower and in the high turrets. The resulting new aloft director was Mk X, in the *California* class. The two tops were occupied by spotters (Spots 1 and 2), with an additional spotter (low spot, or Spot 3) in the fire-control tower, using a separate periscope. The target designator appeared only in the integrated GE Selsyn or Synchro system. Of other early directors (directorscopes), Mk I was a trunnion sight attachment for USSS *Delaware* and *Wyoming*; Mk III was a turret periscope for USS s *Florida*, *Michigan* and *South Carolina*; and Mk IV was another periscope, incorporating a target-bearing transmitter, for the conning tower and superfiring turrets of USSS *California* and *Tennessee*. No Mk V has been identified.

Two of the four triple turrets of USS *Pennsylvania*, probably photographed in December 1916, show characteristic US turret-face sighting ports (below the two wing guns), intended to limit damage from the blast of nearby superfiring guns. All three guns in each turret were mounted in the same slide, because it was US practice to fire full-broadside salvoes. Not visible here are the long-base rangefinders built into the rear of the turrets. The US Navy's Bureau of Ordnance (BuOrd) first proposed a 14in gun in November 1908, but nothing was done at the time. In December 1909 the US naval attaché in London reported that he had no firm knowledge of a British 13.5in gun, but that newspaper reports convinced him that it was under development. In January he reported tests of such a gun. BuOrd then produced curves comparing a 14in/45 with the existing 12in/50. At 12,000 yards the 14in (2600 feet/second) would be slightly less likely to hit (its danger space would be forty-five yards compared to fifty-six for the higher-velocity [2950 feet/second] 12in/50), but it would penetrate more armour (13.8 as opposed to 12.7in). A higher-velocity (2700 feet/second) 14in/45 would be slightly better: danger space would be forty-nine yards, and it would penetrate 14.5in. This improvement was enough for the General Board to recommend adoption of the new gun. On 15 January 1910 BuOrd reported that the prototype 14in/45 had been completed and proof-fired with 'most satisfactory results'.

destroyers. An alternative US-designed Mk 6 performed the same functions.

Until about 1910 it was assumed that a turret should not fire more than one gun at a time, to avoid interference. However, the battleship *Vermont* won annual trophies by firing both guns of each turret together, so that in March 1910 the Bureau of Construction and Repair suggested not only firing all guns together, but also mounting three guns in one turret. Hence the triple turrets in the *Nevada* class, with their single slides for two or three guns. Independent elevation was restored in the next turret design (*Idaho* class), but only to prevent the loss of all three if one jammed (the guns were locked together for firing). Hence the use of a single slide in 8in turrets for the two *Lexington*-class carriers and the early 'Treaty' cruisers (cruisers built under the terms of the Washington Treaty of 1922, which were limited to 10,000 tons and 8in guns). US officers of the 6th Battle Squadron attached to the Grand Fleet during World War I claimed that the British were most impressed by their full-broadside firing (as compared to half-broadsides in British practice). A US officer said that the full salvo gave spotters the best possible view of how shots were falling. Moreover, the denser the salvo, the better the chance that some of the shells would hit.

By 1912 Sperry Gyroscope was developing what it called a fire-control system. It was mainly a data-transmission system using the step-by-step repeater developed for Sperry's gyro-compasses.[15] Sperry's devices were attractive because the existing Cory visuals were proving unsatisfactory. A follow-the-pointer system based on the Sperry transmitter-receiver was adopted for train in 1916 (some ships had it by 1917). Something of this type was necessary if the director was to be used as a target-bearing transmitter. The first follow-the-pointer transmission in elevation was installed on board the battleship *New Mexico* in September 1918. Like its British counterparts, it had to be synchronised (which took time) every time it was turned on, and each transmitter-receiver pair had to be resynchronised every time it was knocked out of adjustment by the shock of shellfire. Switching within the fire-control system, eg, to recover from casualties, was a lengthy process. The stepping motion of the receiver was described as jerky, so it could not be used to control anything directly. By 1918 the BuOrd was looking for a replacement. It chose the synchro (see below). Sperry's transmitters were only the beginning of its interest in fire control. Because its devices could transmit angles, the system included a two-man rangefinder. The US Navy tested it but did not adopt it.[16]

In 1913 the US naval attaché in London was among those fascinated by disclosure of the Pollen fire-control system. He wanted the main US gunnery developers, including the deputy head of the BuOrd, to see it, and he wanted one bought for US experiments. No purchase was made, however, and it is not clear whether any visits took place. However, the attaché's reports (and Pollen's brochures, which he enclosed) seem to have impressed the bureau. In 1914 BuOrd apparently asked Sperry to produce an equivalent to Pollen's system. He was already aware of the Pollen system through the subsidiary he had set up to provide gyro-compasses to the Royal Navy. Sperry's British agent had already recommended licence-production of the Pollen clock.[17] Hannibal C Ford, who designed the battle tracer, left Sperry in 1914 to set up his own Ford Marine Appliance Company the following year. Once that happened, BuOrd asked Ford to design a competing range-keeper, the remaining element of the system (the complete system being a precision rangefinder, a plotter and a range-keeper or clock). In May 1915 Sperry offered a Range Clock and Ford a 'Range and Deflection Predictor'. Ford had already made a proposal.[18] Ford abandoned plans to make gyro-compasses and decided to specialise in fire control, under the changed name of the Ford Instrument Company. His range-keeper was selected after July 1916 trials on board USS *Texas*.[19] The Ford range-keeper was adopted in January 1918 as the primary fleet fire-control device.[20]

The first element of Sperry's system was a true-course automatic plotter – the battle tracer – using a 'bug' that followed a ship's course across a chart. Its plotting arm was activated by a bearing receiver and a range receiver, so it indicated target course relative to the course traced out by the 'bug'.[21] Sperry installed battle tracers in USSs *Utah*, *New York* and *Arkansas* early in 1915. Manoeuvres showed that it was promising but not yet satisfactory, and that it was more useful as a navigational tool than as a way of determining enemy course and speed. Twenty battle tracers were ordered during World War I, and the device was the basis of the later US Dead Reckoning Tracer (DRT). In 1925 British observers reported that all ships plotted own and enemy ships, and that tracking (presumably using DRTs and tracking boards) was widely exercised for gunnery, torpedo control and for tactical purposes.[22]

Pollen's clock may well have inspired Ford, but the two were probably not related in design. Ford was probably the first to separate own- from target-motion. As in Pollen's, the core of the automated system was the Dumaresq equivalent. Unlike a Dumaresq, the US range projector separated own- from target-ship motion, so it was natural for a range-keeper built around it to do likewise.[23] Ford's integrator was an improved version of Pollen's.[24] Ford's device projected range rate along and across, and integrated the rate along to give present range. It also calculated gun range, taking into account target- and own-ship motion while the shell was in the air. Because the range rate across (ie,

USS *Idaho* is shown completing in 1919. This class introduced the 14in/50 gun, carried in separate slides instead of in a single one as in the previous two classes. Guns were still fired together, the US Navy preferring a full-broadside salvo to the British half-salvo (World War II gunnery-training photographs do show partial salvoes, however). The US Navy was proud of its compact triple 14in/50 turret. British constructor Stanley V Goodall, later Director of Naval Construction, was present at first-of-class (*Mississippi*) trials in March 1918. He was impressed by the turrets, but considered them cramped (the main problem was limited headroom). Flash-tightness was taken very seriously, the guns being separated by longitudinal bulkheads and a complete deck separating the turret from the working chamber below. The gunports were small because the trunnions were placed against the face plate, a feature which might cause problems if shells struck that plate (at this stage the guns elevated only to fifteen degrees). Goodall was very impressed by the clear turret roofs, the periscopic sights being placed below the guns in the front of the turret rather than on the roof, where they might be affected by blast (he suspected, however, that the sights of No 1 turret would suffer badly from spray in a head sea). On the other hand, the US Navy had not exercised the care the Royal Navy showed to eliminate rivets inside the turret (in action their heads could shear off and fly about inside the turret). One object of the trials that Goodall witnessed was to measure dispersion (the spread of where shells landed compared to the range at which they were aimed), which was considerable. Goodall gathered from the ship's gunnery officers and from the party from the Bureau of Standards conducting the trials that previous triple turrets had also been subject to undue dispersion. The screened object on the forebridge is a navigational rangefinder, a feature added during World War I (it did not figure in the ship's original plans). It was initially described as a flag rangefinder; when the General Board reviewed bridge arrangements in 1918, it decided that every ship should have both flag and own-ship bridges, the two to be separate. The board assumed that the ship control party would be ten, the flag party three, and the fire-control party eight. More might be needed as fire control developed. The flag or navigational rangefinders became standard on board US battleships and cruisers during the interwar period. They were probably associated in part with the new practice of plotting for situational awareness.

deflection – the amount by which the target had to be led or trailed) was linear (knots) rather than angular (degrees), it could not be integrated to generate bearing. The range-keeper implicitly used target bearing internally, but not in any form accessible to its operators. At a June 1917 Washington fire-control conference Pollen pointed out that dividing the rate across by the range would provide bearing rate, which could be integrated to generate target bearing (Pollen's invitation to the conference suggests his perceived importance at the time). The Ford Range-keeper was modified accordingly, to display generated target bearing on a dial (for comparison with actual target bearing).

Ford's device was much less integrated than Pollen's. It automatically received own-ship course (by gyro),

NAVAL FIREPOWER

New Mexico is shown about 1925. One of her British-type, secondary-battery directors (a small upright cylinder) can be seen on the 01 level abreast her boat crane. This canopy was similar to that used by the Royal Navy. The US Navy characterised these directors as pedestals, and the shields were separately driven. The exception was the *California* and *Maryland* classes, in which these directors occupied windowed spaces in the fire-control tops. In effect the cruiser directors of the interwar period were descended directly from these pedestals. Initially carried in fixed windowed enclosures (as in the *Pensacola* class), they soon reverted to movable enclosures. The ultimate form of such a director was the Mk 34 used by battleships and cruisers to control main armament. The shrouded vertical device abreast her funnel is an anti-aircraft rangefinder. Note the prominent enclosed torpedo-defence platforms on both masts. Periscopes for a director and for fire control (and vision) protrude from the roof of her conning tower and from her B turret. Her foremast carries a clock-like concentration dial, but by this time the US Navy had abandoned British-style turret bearing markings. No forebridge rangefinder is visible (it may have been hidden by the windscreen above the bridge, because it is visible there in an earlier photograph), but one can be seen alongside the catapult atop X (No 3) turret. *New Mexico* reflects the recommendations of the 1919 Fire-Control Board, which called for two separate rangefinders to serve the secondary battery, if possible fore and aft (it treated the navigational rangefinder as one of the two). It envisaged three positions for secondary-battery control: a control station on each mast, well above searchlights and guns; a control position for each group of secondary guns; and a director near the latter for each group. The group control station would include a rangekeeper. It appears that in this ship the screened platform on the foremast was the secondary-battery control position (there was also a platform immediately above it, inside the mast), with the director (in its splinter shield, the small cylinder on the superstructure deck abeam the cage foremast) below it. The mainmast carried a torpedo-control position.

target bearing (from the target-bearing indicator), and stopwatch time. Operators inserted all other relevant data, including initial range and target course and speed (the 'set up', as estimated by plot). They could also insert spots and transmission interval (the estimated dead time between computing and firing, generally about ten seconds). Own-ship speed was inserted manually.[25] Manual input made it easier to integrate the rangekeeper into the existing fire-control system. It was also easier to smooth or filter data before inserting it. Adjustments required more judgement, and the rangekeeper was considered more difficult to use.[26] Moreover, every delay between data creation and insertion made for dead time and hence errors.[27]

Unlike Pollen, Ford was very much aware of errors and the need for feedback. He used horizontal and vertical cross-wire (driven by range and bearing errors) to indicate errors in target speed and course.[28] As in Dreyer's cross-cut, they indicated errors in assumed target course and speed, and thus could be used for correction. Postwar, the British adopted exactly this technique in their AFCT, although during the war they dismissed it as unworkable.[29]

Despite obvious differences, the British gunnery officer sent to the United States after the country entered the war in 1917 found the Ford Range-keeper very similar to Pollen's clock (he mistakenly thought that Ford was the automobile maker!). The wartime US Navy ordered a Pollen clock for trials.[30] Sperry himself believed that Pollen patents had been infringed. When Ford raised this issue, Assistant Secretary of the Navy Franklin D Roosevelt formally guaranteed him against any suits which might be brought.[31] In the 1930s Pollen's engineer Harold Isherwood sued the BuOrd for patent infringement, and the suit dragged on into World War II. BuOrd made the dropping of this suit a condition for Lend-Lease cooperation with the Royal Navy.

Probably due to experience with the British Dreyer Table, a graphic plotter (for rangefinder and gun range and bearing rate) was designed during World War I and added in 1920 to create the standard Mod 3 version of the range-keeper (Mod 2 incorporated the divider, so that it generated bearing). The plotter was a rangefinder data receiver equivalent to parts of the Dreyer Table. However, ranges were entered by hand,

Between the two World Wars the US Navy modernised all its battleships, but early units did not receive improved main-battery fire controls. Thus USS *Arkansas*, shown, had a fire-control top on her foremast similar to that of a *New Mexico* as built. A sketch plan shows a stub tripod abaft her funnel, but it was not installed. No turret rangefinders were installed; instead she was given large units on her turret tops. The catapult atop No 3 turret was also an essential fire-control improvement, as air spotting was expected not only to extend the reach of the battlefleet, but also to make it possible for battleships to fire from behind smoke screens. The new stable verticals were part of the same programme. The stub tripod aft carries searchlights and, below them, a searchlight control position.

New York and *Texas* were the first two battleships to be modernised with the new kind of fire-control system. It comprised Mk XX directors in the cylindrical upper level of each top (Mod 1 in the maintop); a Mk XXI in the fire-control tower (after part of the conning tower), a Mk IX Mod 3 stable vertical in the plot, and a Mk I range-keeper. Apparently there were no directors in the superfiring turret. Unlike the later rebuilt battleships, these two did not have their main-battery elevation increased beyond the original fifteen degrees (corresponding to a range of about 24,000 yards). The United States interpreted the clause in the Washington Treaty prohibiting 'substantial' changes in battleship main batteries to prohibit any such improvement, and in the early 1920s it charged the British with seeking superiority by illegally increasing the elevation of their 13.5in turret guns (which they were not doing). Although a project to increase gun elevation was reported in the late 1930s, it was never carried out. The two high turrets were fitted with rangefinders near their faces instead of near their after ends (note the small protruding ears). USS *New York* is shown, probably in the 1930s. Not visible are the two anti-aircraft rangefinders (vertical rangefinders) on the same level as the main rangefinder, at its after ends. As in *Arkansas*, the stub tripod aft carried searchlights and a control position; during World War II it carried a Mk 50 anti-aircraft director.

The *Pennsylvania* class was unique among US battleships modernised in the interwar period in having an Arma rather than a GE system; BuOrd was determined to maintain competition. Both systems used much the same elements, the Arma system incorporating particular versions. BuOrd standardised its synchros beginning in 1930 so as to ensure that equipment was interchangeable. USS *Arizona* is shown in 1934. Because it employed different synchros, the Arma system required special versions of each component, such as Mk 20 Mod 4 and Mod 5 top directors (both the *New York* and the *Nevada* classes had had the same versions, Mods 2 and 3). Note the double-level conning tower (forward of the bridge) in *Pennsylvania*, which was designated US fleet flagship.

ABOVE The open windows on the upper level of the foretop of USS *Oklahoma* show her Mk XX main-battery director in this 28 April 1931 Puget Sound photograph. As with the cruiser directors, in February 1943 BuOrd approved substitution of a longer-base rangefinder (the stereo Mk 54, 12ft base) for the spotting glass, so that the long-base navigational rangefinder could be eliminated. That seems not to have been done.

RIGHT USS *Louisville* is shown as completed, at Puget Sound on 2 February 1931. Plans originally called for a separate director in a windowed space, with spotting glasses in the top below it. However, ships were all completed with the shielded Mk 24 director shown, incorporating a Mk VII spotting glass. In this system the stable vertical was integrated into the Mk 6 range-keeper, a combination later considered awkward. By 1938 BuOrd wanted to separate range-keeper and stable element, a Mod 8 version of the new range-keeper Mk 8 being developed. It is not certain whether it was installed. The earliest ships of the class were completed without their directors. Originally there were two versions, Mods 0 and 1, the latter being unshielded, in a protected position abaft the after funnel (by 1941 directors were being relocated to the mainmast, just forward of No 3 turret). Operating modes were primary and secondary control. In primary control, the director transmitted gun-elevation orders both to the guns and to the range-keeper in plot. The range-keeper in turn set a mirror in the pointer's sight to the director correction angle (ie, attempted to stabilise the system in line of sight). This method was chosen so that the range-keeper could correct for trunnion tilt by 'weaving the line of sight' up and down rather than by moving the guns. As in other directors, the pointer fired when the target seemed to pass across the cross-hairs in his sight. A similar mirror was mounted in the trainer's sight. Because these sights were more complex than the telescopes of the past, they were both substantial boxes. A sight-setter introduced range into the director for parallax corrections, and also set sight depression for the range and muzzle velocity in use. Range limits were 1000 and 30,000 yards. Apparently there was no cross-leveller, that function being carried out by the Mk 6 range-keeper below decks. In secondary control, the range-keeper in plot trained the director, the trainer applying feedback corrections. Directors were converted to Mods 2 and 3, respectively, in 1941 as part of a larger gunnery modernisation for this class. It entailed installation of an entirely new instrument assembly to measure and transmit director correction, bearing, train designation and gun elevation orders. In this version the range-keeper (presumably Mk 8 rather than Mk 6) no longer transmitted level and cross-level to the director. Director outputs were now the usual ones: director correction (ie, stabilisation in line of sight), target bearing and train designation. This version was connected by shaft to a local auxiliary range-keeper (Mk 7). In secondary control, the director received data from the range-keeper and applied its own trunnion-tilt correction (using a periscope) to the pointer's mirror. Apparently the war intervened, because in 1943 BuOrd was giving instructions to convert not only Mods 2 and 3 but also 0 and 1 to Mods 4 and 5, the main change being to move parallax adjustment so that it referred to a point midway between the two directors.

LEFT The battleship *Idaho* shows her fire-control systems at Norfolk Navy Yard, 3 January 1942. Her Mk 31 Mod 2 main-battery director (the corresponding after unit was Mod 3) seems lost compared with the forebridge 15ft rangefinder and the big Mk 28 anti-aircraft director above and abaft it (which has been partly cut off at the upper edge of the photo). It is surmounted by (and dwarfed by) a Mk 3 fire-control radar (later replaced by Mk 28, a small microwave dish). What look like the arms of a rangefinder are actually those of a Mk VII spotting glass, a stereo device in which each of the observer's eyes is connected to a separate lens. It provided stereo vision but it did not measure relative ranges. In February 1943 BuOrd authorised replacement of the spotting glasses in the battleships and in cruisers with this director with Mk 51 stereo rangefinders (8.2ft base, normally for anti-aircraft), but it is not clear whether this was done. A June 1945 Ordnance Catalogue lists it only on board the battleship *Mississippi* and the heavy cruiser *Minneapolis*. The main-battery system in this class had a maximum range of 36,500 yards (maximum present range in the range-keeper was 36,000 yards); maximum target speed was forty knots, and maximum wind speed was sixty knots. In addition to the two aloft Mk 31 directors, the main-battery system employed a Mk 21 Mod 3 periscope director in the fire-control tower (the after part of the conning tower). In contrast to previous practice, there were no turret directors. The associated stable-vertical director was Mk 30 Mod 1. This was the first system to employ the Mk 8 range-keeper (Mod 2 in this class) standard in World War II battleships and cruisers. A Mk 31 Mod 1 director, at the after end of the bridge structure, served secondary-battery control. The main-battery version had three seats, for spotter, pointer, and trainer. The range-keeper and stable vertical in plot were connected only to the main-battery directors, so the secondary-battery directors had two additional operators, for cross-level and for a Mk 7 Mod 3 range-keeper mounted nearby and connected by a shaft. The brackets on the level above the bridge were for the concentration dial (range dial), which had been removed by this time. In contrast to earlier modernised battleships, in this class the two separate rangefinders were both 15ft-base instruments. At the end of the war the radar antenna mount in this class was modified to take the modern Mk 8 fire-control radar, and it was actually installed only in *Idaho*.

USS *Chester* shows further modifications in this 16 September 1943 photograph taken at Mare Island Navy Yard. Her bridge has been opened to the sky and the Mk 19 anti-aircraft director replaced by the more massive Mk 33, a standard pre-war destroyer type. The old navigational rangefinder is gone altogether, its platform replaced by a pair of 20mm guns. The radar antenna on top of the Mk 24 director is for a Mk 3 fire-control set. Ships of this class had already had their forebridge rangefinders raised to what became the 20mm gun platform level to clear space for enclosed anti-aircraft directors in 1940-41. When BuOrd authorised replacement of the 8in director spotting glass with a rangefinder (to restore the tactical ranging capacity lost when the forebridge unit was eliminated), gunnery officers protested that a rangefinder was by no means equivalent to a spotting glass. With the advent of effective surface-search radar, the tactical rangefinder was no longer vital, but the spotting glass remained an essential back-up (and in some cases supplement) to gunnery radar, particularly before ships were fitted with high-resolution radars like Mk 8.

The heavy cruiser *Portland* shows the 8in director Mk 27 atop her short tripod mast in this 1933 Puget Sound Navy Yard photograph. She and her sister ship *Indianapolis* had Arma fire-control systems. As in the previous *Northampton* class, it was enclosed in a splinter shield (0.5in thick) and had an integral spotting glass (the director 'ears' cover its two lenses). The face of the director has openings (covered here) for the director pointer and trainer. The object on the centreline of the director is presumably the cross-level periscope. Slits in the cylinder below the director mark the 8in control position. The associated stable vertical was the Mk 29 director. Because this stable vertical could not provide cross-level data, Mk 27 incorporated a cross-levelling periscope. The two operating modes, primary and secondary, were defined in terms of reliance on the range-keeper in plot; in secondary mode the system used a local (Mk 7) range-keeper, but still used the fire-control switchboard in plot. These definitions were radically different from those used for the Mk 24 director. The sight-setter's side of the director was used to input range for sight-setting and also to input trunnion-tilt range and deflection (by hand, based on cross-level). Other operators were cross-leveller and the usual pointer and trainer. Like Mk 24, Mk 27 also carried a spotting glass. When facing the target, the trainer was on the left side, the sight-setter on the right, the cross-leveller facing towards the director (away from the target), and the pointer, spotter, and range keeper (operating the auxiliary Mk 7 range-keeper) faced the target. The range-keeper was mounted with its axis along the line of sight to the target, so that its face was visible to both the spotter and the sight-setter. In that way the picture it showed of own and enemy ship could be checked easily against reality. The spotter had a seat above the other operators, so that he could see all of them, and he could also stand to look out through a hatch in the director roof. In this photograph the object covered in canvas behind the windbreak is a Mk 19 anti-aircraft director for the ship's 5in guns. It was served for surface fire by the separate rangefinder just abaft it on the wing of the bridge roof (there was a separate anti-air rangefinder). The similar rangefinder (in the same type of shield) on the forebridge is the usual navigational unit. In February 1943 BuOrd authorised replacement of the spotting glass with a Mk 51 rangefinder, so that the navigational rangefinder could be eliminated altogether.

(continued from page 193)

GE and Ford wanted the new concept applied to the *Northampton*-class cruisers then being ordered. The Bureau was enthusiastic, and the Fire-Control Board agreed; on 18 April 1928 the Secretary of the Navy approved the project. The new light cruisers would have two directors and two computers each.[50] Unlike the *Pensacola*s, they had plotting rooms below decks. The primary mode of control was to couple an aloft Mk 24 director to one of the two Mk VI range-keepers in the plot. The range-keepers generated range and bearing, and incorporated the planned gyro elements (with separate gyros for level and angle) so that they could generate corrections for roll and pitch. As backup, the directors incorporated Mk VII range-keepers, effectively modernised Baby Fords, generating range when pointed at the target.[51] It also controlled 5in secondary guns on battleships.[52] Upon modernisation surviving *Northampton*-class cruisers were fitted with Mk 34 directors and their Mk VI range-keepers were replaced by Mk 8. The successor *Portland* class had Mk 8 range-keepers instead of Mk VI. As in the *Northamptons*, their Mk 27 directors incorporated auxiliary Mk VII range-keepers.

The next approach to consolidation was the combination of a new lightweight Mk 31 director and a new Mk 8 range-keeper in the *San Francisco*-class heavy cruisers and the rebuilt *New Mexico*-class battleships. The director carried a stereo spotting glass, which resembled a rangefinder. Unlike previous directors, it had separate trainer and pointer. Like Mk 6, Mk 8 automatically received ranges from the rangefinders. Like Mk 6, it calculated corrections such as trunnion tilt. Unlike the earlier range-keeper, it did not contain gyros; they were removed to a separate stable vertical communicating with Mk 8. A ship with Mk 8 could fire automatically on level or cross-level, ie, set guns to fire when they were in the appropriate position. It corrected automatically for firing delay. This type of control revealed that synchros sometimes showed time lags, particularly just after being required to reverse as a ship came back from the end of a roll. BuOrd found itself developing a new range of synchros and also experimenting with electronic amplifiers for battleship and cruiser installations.

The Mk 31 director was also used to control secondary batteries in the modernised *New Mexico*-class battleships. Their range-keepers and stable verticals were connected only to their main-battery control systems; in that sense they hardly approached the ideal of consolidation. Thus the Mk 31 secondary directors in these ships needed their own range-keepers (Mk VII) and their own means of cross-levelling (a periscope pointed to the side, visible by its vertical housing). The follow-on cruiser director, Mk 34, was first installed in *Brooklyn*-class light cruisers and in the heavy cruiser *Vincennes* (CA 44). It was heavily modified before and during World War II, with a rangefinder and other features that overloaded it. Even so, it was installed in all the wartime cruiser classes and in battleships modernised in wartime.[53] Because Mk 34 was clearly unsatisfactory, in 1943 it was ordered redesigned as Mk 54, which appeared only in the postwar *Des Moines* class.

The consolidation project was revived in 1935–36 as new battleships were being designed. Now the emphasis was on placing as much as possible of the main-battery system under armour. The main difference between main and secondary battery directors would be the length of their rangefinders. Range-keepers and stable elements would all go below decks. As the initial proposal of 15 November 1935 put it, the resulting system could keep firing, even if the directors were destroyed, because its stable vertical could cancel out roll and pitch and its range-keeper would continue to generate range and bearing. It could even handle enemy manoeuvres if the turrets provided some feedback. This was the closest a battleship could come to eliminating the vulnerability associated with aloft directors.

If the director officer were located outside the director on its platform, the director could be limited to five

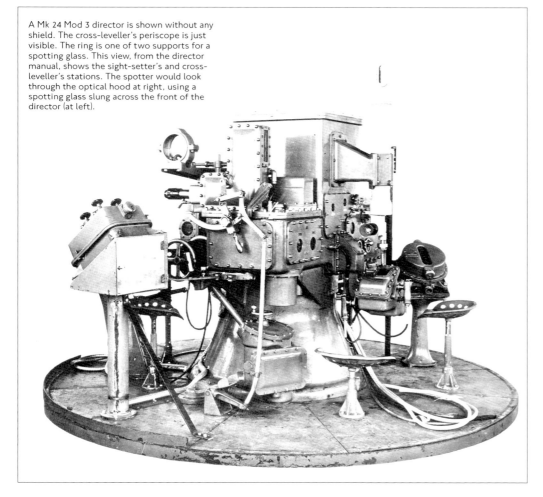

A Mk 24 Mod 3 director is shown without any shield. The cross-leveller's periscope is just visible. The ring is one of two supports for a spotting glass. This view, from the director manual, shows the sight-setter's and cross-leveller's stations. The spotter would look through the optical hood at right, using a spotting glass slung across the front of the director (at left).

ABOVE The cruiser *Northampton* is shown visiting Brisbane sometime between 5 and 10 August 1941. She has been modernised, her after 8in director relocated to the new stub mainmast which is now topped by a 5in anti-aircraft director. The other anti-aircraft director has been relocated to her centreline forward, replacing the navigational rangefinder formerly atop her forebridge (it was relocated to the new mast platform). She also now has a CXAM air-search radar, but as yet no gunnery sets. The boxy anti-aircraft directors are director mount Mk 1, which was built around the existing Mk 19 and the new stereo rangefinder (Mk 42) intended for the new dual-purpose Mk 37 director. It is not clear why it received this designation instead of being considered a new version of Mk 19. Note the bow wave camouflage, intended to confuse a submarine commander looking through a periscope. It would not have had much impact on a gunner using either an inclinometer or the Japanese *sokutekiban* (which was apparently unknown to the US Navy at this time) (see chapter 11). Still at peace, the US Navy had not yet eliminated glass windows subject to blast damage.
(US NAVAL HISTORICAL CENTER PHOTO COURTESY OF EDWARD L O'NEILL)

RIGHT USS *Chester* shows further modifications in this 16 September 1943 photograph. The radar atop her main-battery director is a Mk 3, and the big search radar is an SK. An SG surface-search radar is atop the pole topmast.

operators for a battleship or cruiser, or to four for a destroyer. Even the shield covering the director would shrink. Top-weight would be dramatically reduced. A 17 January memo to Commander Willis A Lee Jr (in charge of the BuOrd Fire-Control desk) described the proposed main-battery system: there would be a single, small aloft spotting position with a target-bearing transmitter, as high as possible (115ft would give a horizon at 25,000 yards) and a topside director (in the fire-control tower) whose main function would be to check the stable vertical. Whether there should be a single stable vertical below decks was not yet certain. In a 27 January memo, Lee differentiated air and surface targets for the secondary battery, so he called for the capacity to engage two anti-aircraft targets using directors on the centreline, and two surface targets using secondary directors lower down on the sides of the ship. He wanted the coming three-turret battleships to be able to engage one surface target (but four-turret battleships should engage two). Heavy cruisers would have battleship capability; large light cruisers would engage two surface and two air targets. By this time anti-aircraft control was so important that there was no longer much point in providing separate because any such a battle would degenerate into a mêlée. Instead, they would intervene only by gunfire, using illumination provided by the destroyers. Formal work to develop night tactics began about 1932, and by 1937 these had become quite sophisticated. One consequence was that US cruisers, but not battleships, practiced main-battery night gunnery (after 1926, battleship night practices were limited to secondary batteries, which would repel destroyer attacks).[4] Despite considerable pre-war practice, much remained to be learned. After the night battle off Guadalcanal, USS *San Francisco* reported that the ship's gun flashes blinded her aloft spotters, a low position for night firing being desirable. The target designation problem had not been addressed pre-war (the ship wanted a fighting bridge with captain and gunnery officer close together).

Pre-war tactics used a mass of destroyers to probe the enemy formation. They would illuminate screening ships for the cruisers to destroy by long-range gunfire. Then the destroyers would go through the breach in the screen to attack the enemy's capital ships. Starshell seemed ideal for such tactics, because it would not necessarily disclose the position of the firing ship, and thus could be used to search an area (there were starshell search procedures). It provided enough light for effective spotting (it was used this way in 1942 night actions). By 1938 the range of 5in starshell was 12,000 yards. Given an effective plot of destroyer positions, a cruiser could engage a considerably more distant target.

Given illumination, firing techniques were not too different from what they were in daylight, using ladders. However, even with starshell, splashes were difficult to spot. Typical hitting rates in peacetime practice were 8 to 11 per cent, far below daytime scores. Effectiveness hinged on volume of fire. The pre-war US Navy thought it had been achieved in the 6in/47 gun, which used cartridge cases to achieve ten rounds per gun per minute. Thus a *Brooklyn*-class cruiser of the late 1930s could fire 150 rounds per minute, like a machine gun. With their bag guns, which fired two or three rounds per minute, heavy cruisers could not produce anything comparable.

The probing concept had little to do with wartime night combat; probably the most important legacies of pre-war thinking were that cruisers were trained to use their guns at night, and that it was expected that they could stay out of torpedo water. Unfortunately, until well into 1943 the US Navy had no idea whatsoever that the Japanese had developed a torpedo – the Type 93 'Long Lance' – whose range was comparable to that of a cruiser gun. Hits on cruisers were generally attributed to Japanese submarines, the explanation being that the US ships had been drawn into a torpedo-submarine trap.[5] There is, moreover, no evidence that the pre-war US Navy was aware of the depth or breadth of Japanese interest in night operations, despite successes in code-breaking that claimed they fully revealed Japanese thinking.

Night combat recalled the problem faced by the Royal Navy before World War I, with the interesting twist that the US Navy did not realise that Japanese torpedo range was comparable to US gun range. There had always been two solutions to the gun-torpedo problem. One was simply to outrange the torpedo; the argument made in chapter 4 is that by 1912–14 the

The Mk 38 director is shown with a late-war Mk 13 radar on top. Note the spotter's periscope (for checking target and general look-out). Normal operating personnel were: spotter, rangefinder operator, stand-by rangefinder operator and talker, pointer, trainer, cross-leveller and radio operator. Total weight, including the antenna of the Mk 8 radar, was 50,000lb, up from the original 40,000lb. The director could train at a rate of ten degrees/second.

British no longer thought they could do so. The other solution was to exploit the low speed of the torpedo, blasting the enemy and then manoeuvering away before his torpedoes could arrive. The British fear before 1914 was that unless they could do one or the other, their battle line would take hits from 'browning shots'. That is exactly what happened to the US Navy in the Solomons. The US cruisers adopted a battle-line formation because it was best adapted to gun fire control and identification (a particular problem under night conditions). Gun flashes from the battle line defined the battle line's course and speed well enough for the Japanese to fire effective 'browning shots'. The situation was worse than that the British had faced before 1914 because, unlike the Germans, the Japanese planned to rely mainly on fire-and-forget torpedoes as opposed to guns that required them to maintain something like a steady course and speed. This greatly reduced any US opportunity to destroy the Japanese ships before their Long Lance torpedoes arrived. Note that night conditions around Guadalcanal resembled those the British expected in the North Sea before World War I: visibility was typically less than 10,000 yards, sometimes less than 5,000 yards.

In pre-war and early wartime practice, the situational awareness of a force was centred on the plot aboard the flagship. Given a valid plot, the force commander could issue appropriate orders for ships to

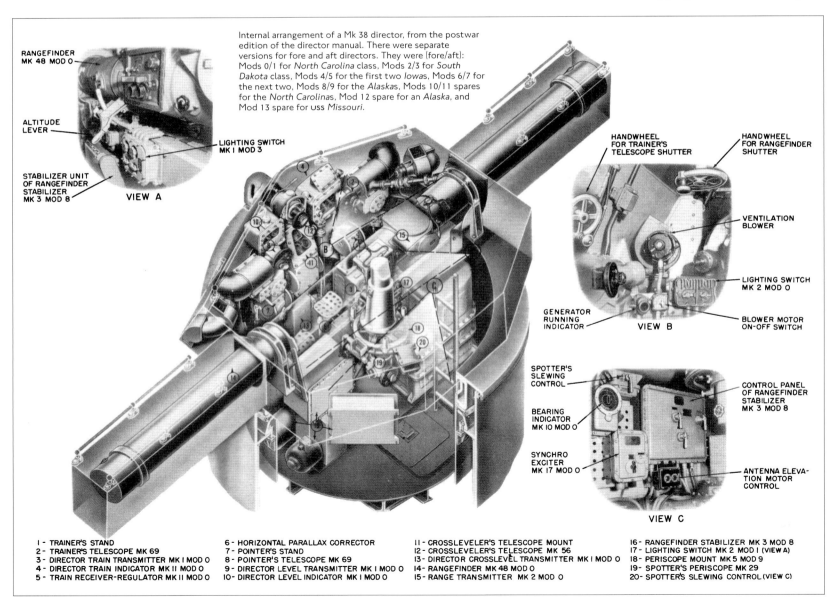

Internal arrangement of a Mk 38 director, from the postwar edition of the director manual. There were separate versions for fore and aft directors. They were (fore/aft): Mods 0/1 for *North Carolina* class, Mods 2/3 for *South Dakota* class, Mods 4/5 for the first two *Iowas*, Mods 6/7 for the next two, Mods 8/9 for the *Alaskas*, Mods 10/11 spares for the *North Carolinas*, Mod 12 spare for an *Alaska*, and Mod 13 spare for USS *Missouri*.

ABOVE: Main-battery plots were sufficiently complicated – and crowded – to warrant building mock-ups. This is the main-battery plot of USS *North Carolina*, the first of the new battleships, in mock-up form (looking forward) at the Brooklyn Navy Yard, 5 June 1940. The large objects are her two Mk 8 range-keepers, with their graphic plotters (note the stool for the graphic-plotter operator). To the right is the fire-control switchboard.

BELOW: Stable vertical Mk 41 (foreground) and range-keeper Mk 8 Mod 9 in the main-battery plot of USS *North Carolina*, 6 February 1942. The ship's fire-control switchboard is in the background.

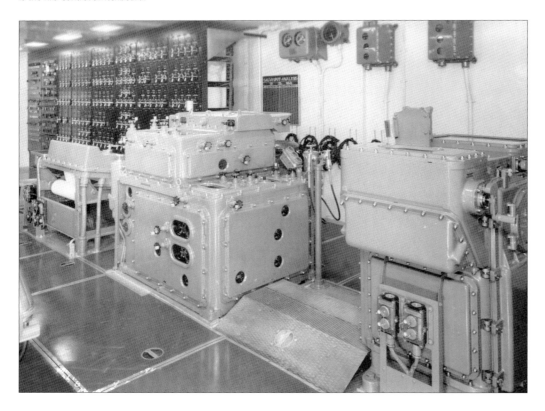

manoeuvre and fire. In theory the line-ahead formation simplified the problem, because the commander automatically had a reasonable idea of where his own ships were, and did not need to fear friendly fire. Automatic plots (in the US case, dead reckoning tracers [DRTs]) were essential to this process. They helped individual ships keep track of their own positions, so that they could usefully report contacts relative to those positions. An unpleasant early-war surprise was that the 'bugs' used to keep track of own-ship position in a DRT tended to jump off the plotting board at the shock of gunfire. Individual ships did not maintain an overall plot, but relied on positive commands from the flagship. The night battle off Guadalcanal on 12–13 November 1942 showed the pre-war system at its worst. The flagship *San Francisco* lacked surface-search radar, so its plot depended on what the ship with such a radar (*Helena*) saw *and reported*.[6] The Japanese adopted a confusing three-division formation (two flanking forces advanced forward of the main body), which could be seen clearly on radar but which may not have been nearly so clear on the flag plot. Exactly what happened is unknown, because Japanese fire wiped out everyone involved aboard the flagship.

Concentrating situational awareness on the flagship made for sluggish reactions by the other ships in the force. That was acceptable in attacks on a Japanese battle line within its screen, its ships moving carefully to avoid collisions. But it was understood that the system would break down in a confused tactical situation, which was precisely why pre-war practice kept the cruisers outside the torpedo battle within the enemy screen. One reason the system inevitably broke down was that there was no way to send (or, for that matter, to absorb) enough information to describe a rapidly changing tactical situation. The system worked pre-war because destroyer and cruiser divisions were permanent, and spent the training year working together. Their commanders shared a mutual understanding of their reactions. Unfortunately, wartime demands broke up the formations. As a consequence, battles were often fought by commanders who did not know how to react during a fast night battle. That applied to the two night

The new battleships all had Mk 40 auxiliary directors in their conning towers. The early version, in the *North Carolina* and *South Dakota* classes, carried a Mk 9 spotting glass with a 15ft baseline (the entire tube was about 16ft long and 13in in diameter). It is visible here atop the conning tower of USS *South Dakota*, off Puget Sound Navy Yard on 24 August 1944. Abaft it is the ship's forward Mk 37 dual-purpose (5in/38) director; the principal main-battery director is the Mk 38 atop the tower foremast. The Mk 40 director also used two Mk 30 periscopes, for its leveller and cross-leveller (they are barely visible here). Beside them and the spotter, it employed a computer operator (for a Mk 3 computer), a range talker, and follow-up operators for sight and sight deflection, level and cross-level. In the *Iowa* class the spotting glass was replaced by a two-operator Mk 32 periscope, and a Mk 3 (later Mk 27) radar was added. Since it was not stabilised, Mk 40 could not be used for blind fire. Mk 55 was the analogous conning-tower director in *Fargo*-class light cruisers.

Plans for the new World War II battleships all included a pair of short-base (12ft Mk 35) rangefinders, one on the forward superstructure and one atop No 3 turret. Only the *North Carolina*s were so fitted. USS *North Carolina* is shown, newly painted in camouflage instead of in the previous two tones, at New York Navy Yard in November 1941. Note the rangefinder atop her pilot house and the Mk 9 spotting glass atop her conning tower (which included her fire-control tower). By this time the second Mk 35 atop No 3 turret had been removed; the forward rangefinder followed during 1942. Plans also called for concentration dials like those of earlier battleships, but they were never fitted.

battles of Guadalcanal in November 1942. The wartime Royal Navy found itself in much the same situation.

The navy much preferred control by exception: to proceed unless ordered otherwise. That in turn required each ship commander to know the tactical situation. In poor night visibility that seemed impossible. The single central commander therefore took responsibility for understanding the overall tactical situation, issuing orders based on that understanding. Much the same consideration applied to a fleet spread out over a wide area, parts operating beyond the horizon of any single ship. This type of centralised control carried risks. The central commander relied on reports from the dispersed ships. The picture he used took time to assemble, and was subject to error because, for example, units might not always report correctly. Orders based on the picture aboard the flagship might not quite correspond to reality. The faster the action, the worse the problem. Tactics were designed to reduce the burden on the commander maintaining a tactical picture. That was one reason for adopting line-ahead formations: anything to either side of the line could be considered hostile, and could be engaged without requiring permission.

The logic of central command was much better suited to a gunnery battle, in which damage would be cumulative, than to a torpedo battle in which attacks could be sudden and catastrophic. Moreover, given the relatively short range of American torpedoes, ships delivering them had to leave the line to approach the enemy, thus greatly complicating the task of distinguishing friend from foe. Destroyers, for example, were safe from their own side's fire only as long as they remained in the line-ahead formation. When they left to deliver torpedoes, they could be (and were) mistaken for the enemy. When the line kinked, as it did off Guadalcanal on the night of 12–13 November 1942, ships that had been in line suddenly appeared in the free-fire zone to one side. They were promptly engaged and, in some cases, sunk. The pre-war US Navy was aware of the problem, and it had coloured fighting lights for night identification. Unfortunately these lights provided the alert Japanese with an aim point, hence could not be used.[7] The solution was electronic identification (IFF, identification friend or foe) installed with ships' radars.

The night battle off Guadalcanal demonstrated the limitations of the pre-war centralised command system. Only one ship in the force, the cruiser *Helena* (not the flagship) had an effective surface search radar (an SG). Admiral Scott commanded from the cruiser *San Francisco*.

In accord with pre-war practice, he based his tactical decisions on a plot his ship maintained. In this case its most effective sensor was on *Helena*, and the

The Iowas were the culmination of US battleship design. Although nominally less powerful than the Japanese Yamatos, they used super-heavy shells which would have been at least as damaging at long range. They also had much better fire-control systems, particularly their radars. Against this, the Iowas spent World War II as escorts for fast carriers, with limited opportunities for main-battery gunnery training. Crews lost proficiency. Specifications for the main-battery fire-control system issued in September 1940 required the system to handle present ranges out to 50,000 yards, with a maximum roll angle of twenty degrees (period sixteen seconds), a maximum pitch angle of five degree (period eight seconds), and a maximum yaw of 1.5 degrees (period eight seconds). Maximum range spots were down 2000 yards and up 3000 yards. Own ship speed could be up to thirty-five knots, and target speed up to forty-five knots; wind speed could be up to sixty knots. The system had to handle gun-velocity losses of up to 300 feet/second; initial velocity for the 2700lb shell was 2500 feet/second (the other nominal velocity was 1800 feet/second). Plans called for installing concentration dials, but neither they nor their supports were ever fitted. Each turret had an auxiliary main-battery computer (Mk 3 Mod 2) in the booth at its rear, for local or alternative control. Range was a hand input. The Iowas were the first US battleship class with two plotting rooms, hence had the most invulnerable main-battery system of their time (the Alaska class had a similar system, but they lacked a Mk 40 director, and the postwar Des Moines-class heavy cruisers also had two plotting rooms). USS Wisconsin is shown here preparing for the Jamestown International Naval Review, 12 June 1957. A year later she was placed in reserve, the last active US battleship – until New Jersey was revived for Vietnam and all four Iowas were revived in the 1980s to form the cores of Surface Action Groups which could supplement the carrier battle groups, allowing the US Navy to spread itself more widely against the Soviets and their surrogates. In both cases the battleship was attractive because, unlike an air force, she could bombard with little danger. During the Vietnam War the Naval Ordnance Systems Command designed sub-calibre rounds, which a battleship like New Jersey could use to hit any target in North Vietnam. The ship was placed in reserve before they could be built. Reportedly the North Vietnamese made it a condition of the peace talks begun in 1968 that New Jersey cease firing at their territory. The Iowas were retired again largely because of fears that the 1991 accident to No 2 turret on board Iowa might be repeated. With the Cold War over, it no longer seemed so vital to be able to cover many stations simultaneously – but that requirement seemed to be returning after 9/11. Wisconsin is currently open to visitors in Norfolk, Virginia.

The first great change for US battleships after the outbreak of World War II was radar. USS Colorado is shown at Puget Sound, 9 February 1942, with a Mk 3 fire-control radar, effectively a radar rangefinder, atop her foretop. Otherwise the main changes are extensive splinter protection for her anti-aircraft battery and numerous 20mm guns. The ship in the background is her sister Maryland. Both ships have been fitted for, but not yet with, air-search radar antennas (which would go on their stub foretopmasts). Because neither ship had been damaged at Pearl Harbor, neither was ever rebuilt during the war. Even so, both were retained in reserve until 1959 as potential shore-bombardment ships.

information he received from her was not always either correct or consistent (his radio log showed several attempts to clarify what he considered illogical information).[8] Even if the cruiser's information had been delivered perfectly, using another ship's radar as the primary sensor for tactical awareness would have imposed unacceptable delays in a fast-moving situation.

The situation changed radically when all ships were fitted not only with effective surface-search radars but with Combat Information Centers (CICs), in effect local plots which provided each ship with sufficient situational awareness to enable it to operate even at night with considerable autonomy. In pre-CIC days, it would have been impossible for each ship to maintain an accurate plot of group movements at night, simply because it would have been impossible for each to send and receive sufficient numbers of messages from other ships reporting their positions and observations. The radar feeding a ship's CIC did much of this work. Now the plotting team had to associate series of radar detections to form the tracks of various friendly and enemy ships, and then to sort them out. Ship-to-ship messages were reduced to those indicating track identification, a load which any ship could handle. Tactics could become much more fluid. One reason was that ships could function more autonomously, more the way they might act in daylight. Another was that the cycle of observation and decision-making was dramatically shortened. Moreover, unlike searchlights, surface-search radar could detect targets at night without giving away a ship's presence. A surface force using radar and CICs could maintain the element of surprise until its guns began to fire.

However, the potential of a CIC-equipped force to use flexible tactics was not realised until well into 1943, because such tactics did not seem warranted until the full potential of the 'Long Lance' was understood. At Cape Esperance (11–12 October 1942) the US force crossed the Japanese 'T' and almost immediately knocked out two of the three Japanese cruisers. As Jellicoe might have said, the line-ahead formation favoured long-range gunnery over shorter-range torpedoes; destroyers had to run in to use their torpedoes. When that happened, they were subject to accidental attack by friendly fire. Thus the destroyer Duncan was hit during her torpedo run.

The battleships damaged at Pearl Harbor were largely rebuilt. *West Virginia* shows just how much could be done. She was fitted with an all-dual-purpose secondary battery and with a fully modernised main-battery control system, including remote control. Space and weight precluded installation of the most modern battleship system, so she (and other old battleships) received a cruiser director (Mk 34), surmounted by a Mk 8 fire-control radar. In this form she proved her value at Surigao Strait in October 1944. Of the surviving old battleships, only *West Virginia* and the two *California*s received remote control for their main-battery turrets, making it possible for them to move to cancel out both roll and cross-roll. *West Virginia* is shown on 2 July 1944, her modernisation just completed. At Surigao Strait, she straddled on her first salvo, as the US Navy had hoped to do before World War II, and thereafter kept hitting, with a few changes in range but none in deflection. In their article on US World War II battleship gunnery, Fischer and Jurens consider this performance the best of any battleship in World War II, confirmation of the efficacy of the new US radars and computers. (See 'Fast Battleship Gunnery in World War II: A Gunnery Revolution,' by B D Fischer and W J Jurens, *Warship International* Vol 42 No. 2 and Vol 43 No. 1.) By way of contrast, *California* began about 800 yards over, perhaps reflecting overshooting. She straddled on her third and all subsequent salvoes.

A month later, however, off Guadalcanal (12–13 November) the US line broke up and the battle became confused. When the light cruiser *Atlanta* turned out of line, flagship *San Francisco*, astern of her hit *Atlanta* with nineteen 8in shells at a range of about 3000 yards.[9] *Atlanta* may have been in the van to back up the destroyers leading the force, should they be detached for torpedo attacks, and because her 5in guns could fire starshell.[10] On the other hand, the US force managed to get close to the Japanese main body, one of whose two battleships, *Hiei*, was so badly damaged by 8in fire from *San Francisco* and *Portland* that she was found adrift by bombers the next day and sunk.[11] For her part, *San Francisco* survived because the Japanese ships were en route to shell Henderson Field on Guadalcanal, and were therefore armed with shore-bombardment shells.[12] A 5.5in shell from the battleship *Hiei* hit the signal bridge of *San Francisco*, killing force commander Rear Admiral Daniel J Callaghan and his staff. According to the *San Francisco* after-action report, towards the end of the action it appeared that the two Japanese columns were firing at each other. One welcome conclusion from the battle, as reflected in another after-action report, was that the Japanese were not using any sort of fire-control radar. They were relying heavily on searchlights for fire control. The US Navy should therefore try to fight at ranges at which Japanese searchlights would be useless.

In the next cruiser battle (Tassafaronga, 30 November 1942), the line of US cruisers ran at high speed (to avoid the supposed submarine torpedo threat) while engaging Japanese destroyers at long range with their guns. Their muzzle flashes gave away their positions, much as Callaghan seems to have feared. The long-range Japanese torpedoes were ideal night weapons because, in effect, they had very long danger spaces, and thus did not demand precise range data. Four US cruisers were hit, one (*Minneapolis*) by two torpedoes; *Northampton* was sunk.

The significance of the Long Lance was still not recognised eight months later, when the same tactics led to the loss of the light cruiser *Helena* at Kula Gulf

When US battleships were modernised during World War II, in most cases priority went to their anti-aircraft battery. *Nevada* retained the small circular platforms for her pre-war Mk 20 main-battery directors, though not the control platforms below them. Not until Surigao Strait (October 1944) was it clear that the pre-war system was far less effective than the later one built around a Mk 8 range-keeper and a Mk 34 or 38 director. The rangefinding radars above the main-battery directors are Mk 3s. The large air-search radar on the foremast is an SK; the mainmast carries the smaller SC-1. *Nevada* is shown here on 2 September 1943.

units being produced between December 1943 and April 1945.

As in the Royal Navy, considerable effort went into measuring target inclination. All modernised battleships and heavy cruisers except the *Mogami* class used a separate device, called a *sokutekiban* like the earlier Dumaresq-equivalent but far more sophisticated. It measured target inclination both directly (by inclinometer) and indirectly, by observing how target range and bearing changed over time. For the latter, it used an optical technique developed by Barr & Stroud to solve a pair of equations linking target bearings and ranges at two separate times with own-ship speed and course. The solution gave target speed and course. Separate operators performed each function: following up the ship's gyro-compass (ie, entering heading); measuring enemy change of bearing; training the *sokutekiban*; entering target length and range differential; following-up (entering) measured inclinometer angle; following-up (entering) present range; calculating target inclination; and transmitting target speed and inclination down to the computer.

Type 92, which equipped all the modernised battleships, had a maximum measured range of 40,000 metres (43,744 yards) (maximum firing range was 39,800 metres/43,525 yards). Maximum target deflection was 130 mils right and 160 mils left. Maximum own-ship speed was thirty knots (enemy ship speed was forty knots, so maximum change of range rate was seventy knots). The system could handle winds of up to thirty metres/second (108 kilometres/hour, or about fifty knots).

Yamato had the latest system, comprising the Type 98 table, Type 98 director, and Type 98 *sokutekiban*. As in the *Mogami*s, the *sokutekiban* was an appendage to the table, receiving data from the director. Unlike its predecessors this was a series rather than a reciprocal system, the table transmitting train and elevation directly to the guns, as in western systems. This system also introduced automatic (electro-mechanical) follow-ups, which had been lacking in earlier systems. Measured range limit was 50,000 metres (54,680 yards); gun-range limit was 41,300 metres (45,166 yards). Own speed could be up to thirty-five knots, enemy speed up to forty knots, and wind speed up to forty metres (forty-four yards)/second (about seventy-eight knots). Deflection limits were as in *Nagato*.

Ships generally had multiple directors, but their electrical systems made switching between directors difficult, as most systems were on single selsyn circuits. To switch, both directors had to be turned to either ninety- or 270-degree bearing, as were the guns; all guns had to be set to one elevation (ten degrees), and

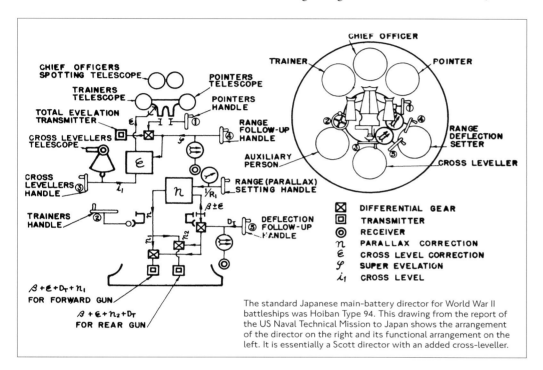

The standard Japanese main-battery director for World War II battleships was Hoiban Type 94. This drawing from the report of the US Naval Technical Mission to Japan shows the arrangement of the director on the right and its functional arrangement on the left. It is essentially a Scott director with an added cross-leveller.

The most unusual feature of Japanese heavy-calibre fire-control systems was a special device, the *sokutekiban*, intended to measure target course and speed. In effect it supplemented or superseded an inclinometer. This photograph is from the US Naval Technical Mission to Japan report.

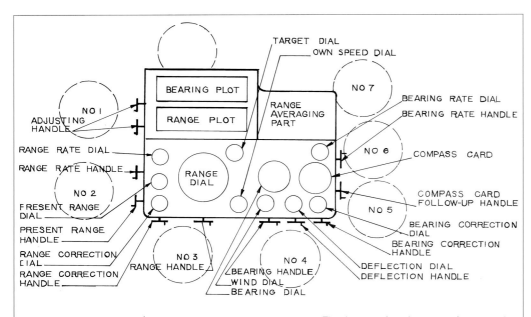

The Type 92 low-angle table (ie, analogue gunnery computer) equipped most Japanese battleships (Type 94 was the corresponding secondary-battery table). Many of the handles shown here were used to repeat results of computation in one part of the table so that another part of the table could use them. The circles indicate the operators. This drawing is based on one in the report of the US Naval Technical Mission to Japan (language in the report suggests that it was written by British gunnery officers).
(A D BAKER III)

This photograph of the partly disassembled Type 92 table (fire-control computer) aboard the battleship *Nagato*, taken just before the Bikini atomic-bomb test, gives some idea of how complex such devices were. The Barr & Stroud system and its derivatives were far more compact than (and far less sophisticated than) the contemporary Admiralty Fire-Control Tables (AFCTs).

all range dials, including the computer, had to be put to one setting (10,000 metres/10,936 yards); all deflections had to be zeroed. Only then could switching be done. Under the best conditions this would take at least a minute, and switching was needed only under bad battle conditions. A postwar Allied report added that Japanese switchboards were poorly arranged, with switches badly scattered, further complicating practice.[9]

The writers of a postwar Allied report on Japanese naval fire control were impressed by Japanese insistence on very tight salvoes, which had been used to good effect at Leyte Gulf (presumably in the Battle off Samar). They felt that to some extent it offset their lack of effective radar. Salvo tightness was due in large part to a special Type 98 (ie, 1938 [see note 3 for an explanation of Japanese deisgnations]) trigger-time-limiting device in the transmitting station (plot). It reduced the firing interval once the circuit had been closed (it

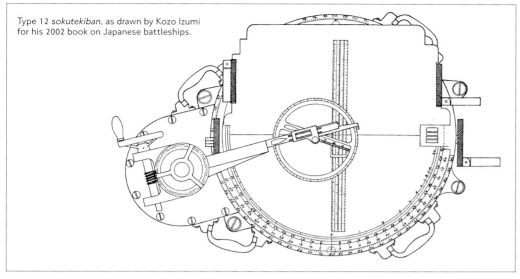

Type 12 *sokutekiban*, as drawn by Kozo Izumi for his 2002 book on Japanese battleships.

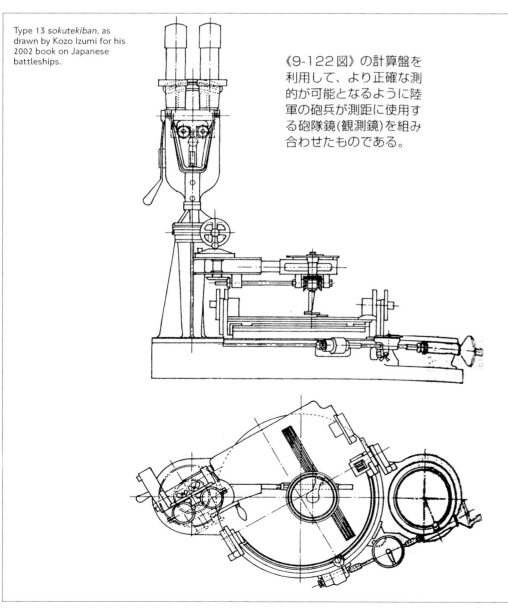

Type 13 *sokutekiban*, as drawn by Kozo Izumi for his 2002 book on Japanese battleships.

《9-122図》の計算盤を利用して、より正確な測的が可能となるように陸軍の砲兵が測距に使用する砲隊鏡(観測鏡)を組み合わせたものである。

was adjustable between 0.08 and 0.2 seconds). There were also Type 98 delay coils to keep guns in the same turret from interfering with each other. On the other hand, Japanese gyros and stable verticals were considered vastly inferior to those of the Allies. The Japanese had made enormous efforts to develop effective night optics, but it seems that the lack of gyros would have handicapped them badly in night gunnery.

Work on a gyro horizon (stable vertical) only began in 1932, the prototype being tested in 1935 on board the gunnery trials ship *Hiei*; it was adopted in 1938 as Type 98. Its worst problem was that its follow-up could not follow the stable vertical smoothly enough. An improved Type 1 Gyro Horizon was tested on board the carrier *Shinyo*. At the end of the war this was only an experimental device.

Except for anti-aircraft control, the Japanese used coincidence rangefinders.[10]

Until the mid-1920s Japanese 355mm (14in) and 406mm (16in) shells were Type 3 (ie, 1914) CPC (see Appendix) derived from the British shells supplied in 1913 with the battlecruiser *Kongo*. Like the CPC used during the Russo-Japanese War, they were filled with Shimose explosive, which tended to explode on contact (like wartime British Lyddite CPC).[11] The Japanese were impressed with the wartime German use of AP shells at long range, and they obtained some. The first of a new series of APC shells was tested in 1924 against the incomplete new battleship *Tosa* and the old *Aki* and *Satsuma*; a new No. 5 APC shell was adopted in June 1925.

The trials unexpectedly showed that the new shells could do considerable damage when they hit underwater. In June 1924 a shell fired at 20,000m (21,872 yards) hit the water twenty-five metres short of *Tosa* and continued to hit the ship underwater. The torpedo protection had no effect whatever, and the shell opened a large hole in a boiler room, admitting 3000 tons of water. A similar hit sank *Aki* on 6 September 1924 when it flooded her engine rooms. These experiences led to extensive work on shells that would follow a consistent underwater trajectory. The Japanese decided that allowing for such hits would more than double the

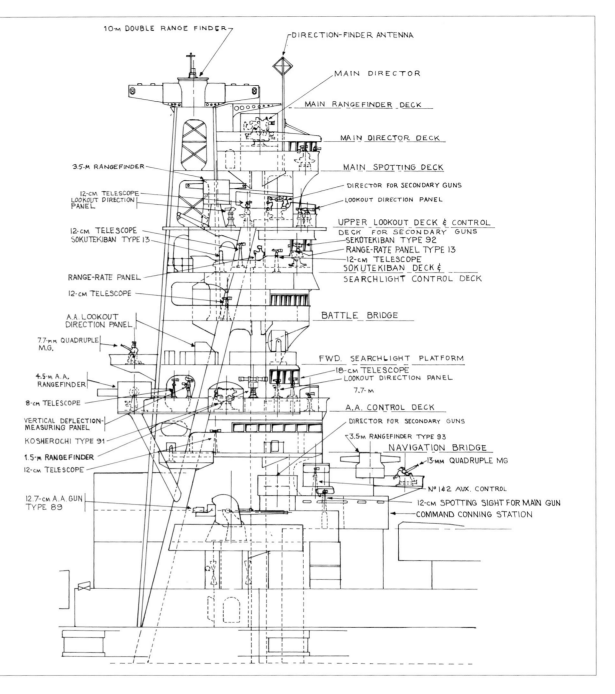

The Imperial Japanese navy conducted a more thorough battleship reconstruction programme than any fleet except the Italian, extending ships' hulls, adding considerable power, and modernising fire control (with added gun elevation to match). The two *Fuso*-class battleships (*Yamashiro* is shown) were the first, and their 'pagoda' foremasts were the most extreme the Japanese fitted. Note that the 'pagoda' was built around the ship's original tripod foremast. In the 1920s these ships added numerous platforms to their masts, but the 'pagoda' represented an attempt to integrate them properly. The complexity of the mast could be attributed in part to the insistence not to combine functions. Thus the Japanese continued to use separate directors and rangefinders, as in World War I British practice. They benefited less than they should have from having multiple directors because their fire-control switchboards and data transmitters did not permit quick switching from one director to another. Note the *sokutekiban* level, with its separate instrument and range-rate panel. Note too the separation between the battle bridge and the navigating bridge (or compass platform, in British parlance). The device marked 'Kosherochi Type 91' is Kosha Sochi Type 91, a high-angle director introduced in 1931. Like the contemporary US Mk 19 (but unlike contemporary British AA directors) it did not incorporate a rangefinder (note the separate 4.5m (14.7ft) rangefinder on this level). The penalty, which was deadly in wartime, was that the rangefinder was not always focused on the same target as the director. The presence of the 3.5m (11.5ft) navigational rangefinder suggests that the Imperial Japanese navy had learned to use plots to maintain the situational awareness needed to execute the complicated tactics it espoused. This drawing was adapted from one in the fire-control report of the US Naval Technical Mission to Japan.

(A D BAKER III)

effective height of the target and thus greatly increase its danger space at very long range.

The first shells designed for underwater attack were Type 88 (ie, 1928) or No 6, tested in 1927 and formally adopted on 17 November 1928 for 203mm (8in), 356mm (14in), and 406mm (16in) guns. They were superseded by Type 91 (ie, 1931), the World War II standard. In both types, the streamlined windshield broke away on impact with the water, leaving a flat head stable underwater. Although it would lose velocity quickly, such a shell would retain enough for a short distance (typically half its initial velocity after travelling 100 calibres from the point of impact with the water, and one-tenth after travelling 200 calibres). In addition to its underwater capability, Type 91 was boat-tailed (with the after end tapered down) for greater range. In 1941 the Japanese introduced dye loads for battleship shells (which were designated Type 1) so that splashes from different ships could be distinguished; 203mm (8in) cruiser shells were similarly

The gunnery itself was not too impressive. In the Java Sea *Nachi* and *Haguro* fired 1271 shells at three Allied cruisers, but had only five hits (of which four were duds) at ranges of 20,000 to 25,000 metres (21,872 to 27,340 yards) (cruiser torpedoes sank two Dutch cruisers and a Dutch destroyer). Later four heavy cruisers fired 1459 shells at the British cruiser *Exeter*, the British destroyer *Encounter*, and the US destroyer *Pope*, opening at 23,000 metres (25,150 yards). *Exeter* went unscathed for an hour, and was sunk only after a boiler-room hit slowed her. *Encounter* was also hit, but *Pope* escaped (only to be sunk by aircraft). On another occasion two Japanese heavy cruisers fired 170 shells at about 5300-metre (5796-yard) range (using starshell illumination) to sink the US destroyer *Pillsbury*. In the night action at Savo a few months later, the Japanese cruisers managed about 10 per cent hits at 5000-metre (5468-yard) range. Although the targets of these cruisers were also hit by torpedoes, gunnery was apparently decisive in this action. At Cape Esperance US radar gunfire hit two Japanese cruisers before they could do much damage. The other two Japanese cruisers hit the US cruisers *Boise* and *Salt Lake City* at ranges of 8000 and 7400 metres (8750 and 8090 yards), respectively, apparently using searchlights on *Boise* as aim points. They made eight and two hits, respectively, the former including the only success by a Type 91 shell during the Pacific War. During the night battle of Guadalcanal two heavy cruisers inflicted twenty-one 203mm (8in) hits on the battleship *South Dakota* at 5000-metre (5800-yard) range. The next gun action was the inconclusive Battle of the Komandorski Islands (see chapter 10, page 225), and the final one was the Battle off Samar, in which a large Japanese surface force sank a small part of a much weaker US force of escort carriers, destroyers and destroyer escorts. Most of the US destroyers and destroyer escorts survived by 'chasing splashes', which meant manoeuvering inside the spotting-correction loop of the Japanese fire-control systems. It is not clear to what extent the success of such manoeuvres indicates that the Japanese system was more sluggish than others.

The main wartime surface-gunnery development was radar. Apparently it offered very little in the way of situational awareness and definition (at least in bearing). These limitations seem to have been demonstrated at Surigao Strait, where the Japanese force was unable to defend itself.

The light cruiser *Sakawa* illustrates the standard late-war Japanese radars in this postwar photograph. Her main-battery director carries the 'mattress' of Type 2 (ie, 1942) Mk 2 Mod 1. The object on the main mast is the Type 3 Mk 1 Mod 3 air-search set. Just below and alongside the director are the paired horns (with another pair on the other side) of the Mk 2 Mod 2 Kai-4 surface-search radar. None of these radars seems to have been intended specifically for gunnery purposes. Note, for example, that the mattress was not mounted directly on the director (and, in any case, that the Japanese tended not to mount their rangefinders on their directors). Work on Type 2 Mk 1 began in October 1941. It used a 4 x 3 mattress and operated at P-band (1.5m/4.9ft wavelength) and at a peak power of 5kW, perhaps 1 per cent of the power of contemporary US and British sets operating at this frequency. Range was seventy to 100 kilometres, accuracy was one to two kilometres (resolution two kilometres and twenty degrees); the antenna measured 6 x 2 metres. The air-search radar operated at a similar frequency. Derived from a land-based set, its development was completed in February 1944. Peak power was 10kW, and range was fifty to 100 kilometres. The horns indicated that Mk 2 Mod 2 (unusually, for a Japanese device, with no Type number, because this was an experimental designation) operated at microwave frequency (10cm, ie, S-band). Development was completed in December 1943. Peak power was 2kW, again only a very small fraction of what was common in the US and Royal Navies. It could detect a battleship at twenty-five kilometres (thirteen nautical miles). Although designed from the outset for naval use, it was not considered as reliable as the simpler metric-wave types. It required two horns on each side because it had separate transmitting and receiving elements. The beam was 38 x 38 degrees. Given their limitations, none of these radars could be considered even moderately effective for gunnery. This ship was armed with a 15cm/5.9in/50 Type 41 gun derived from a Vickers gun used as a secondary battery in *Kongo*-class battlecruisers and then in the *Fuso* class. These weapons were actually of 152mm (6in) calibre, as might be expected of guns designed and built in the United Kingdom. They fired a 100lb shell at 850 metres/second (2788 feet/second). However, the roughly contemporary *Oyodo* class was armed with triple 155mm (6.1in)/60 mounts which had been removed from the *Mogami*s when the latter were rearmed as heavy cruisers (55.87kg/123 lb shell fired at 920 metres (3018 feet)/second). For comparison, the 14cm/50 Type 3 gun that armed Japanese light cruisers fired a 38kg (83.7lb) shell at 850 metres (2788 feet)/second. It had been designed specifically to replace the 150mm (5.9in) gun, whose shell was considered too heavy for Japanese seamen. This weapon armed *Ise*- and *Nagato*-class battleships.

CHAPTER 12
The French Navy

Paris is shown at Plymouth, England in 1941. Damaged by bombing on 11 June 1940, she had sailed to Plymouth for repairs. There she was seized by the Free French upon the fall of France. During the war, she served as a base ship for small craft, returning to France after the war. She survived as a base ship at Brest, being stricken on 21 December 1955 and broken up in 1956. The cruiser-type director visible here was installed during an August 1927–June 1929 refit at Toulon (the 340mm/13.4in-gun ships had different directors). It carried a 4.5m (14.7ft) coincidence rangefinder and a 3m (9.8ft) stereo rangefinder; as in the cruisers, the latter was mainly for ecartometry (measurement of the vector between splash and target – for correction of fire). At that time the old triplex rangefinder atop the conning tower was replaced by a duplex unit carrying two 4.5m (14.7ft) coincidence rangefinders. Atop B turret and atop the now-duplex rangefinder (on the conning tower) were anti-aircraft rangefinders (1.5m/4.9ft stereo). Another 4.5m (14.7ft) rangefinder was added aft. Also added were two secondary-battery directors, on the navigation bridge, carrying 2m (6.5ft) coincidence rangefinders. This was also when concentration dials were fitted to the front and sides of the foretop. Early in 1930 the big rangefinder atop B turret (8.2m/26.9ft coincidence unit) replaced one of the anti-aircraft rangefinders. By the early 1930s these ships were considered obsolescent, so in 1933 plans to fit them with anti-aircraft directors were cancelled. *Paris* could be distinguished from the other two ships of this class because her two forefunnels were not enclosed in a single housing. She became a school ship for electrical and torpedo ratings in 1932, then for boilers in 1936. Of the other two ships of the class, *Courbet* (gunnery training ship from 1931) was also in England when France fell. She was scuttled on 9 June 1944 as part of the artificial port off Normandy. *Jean Bart*, renamed *Ocean* in 1937 when the new battleship of that name was laid down, became an electrical and torpedo school ship in 1935, then an electrical and radio school ship in 1936. She was decommissioned on 1 December 1937, the other two ships forming the French 3rd Battle Division in June 1939 (but soon reverting to school duties). Because she had been disarmed at Toulon, *Ocean* was not scuttled at Toulon, but was used by the Germans as a target, then hit by a US bomb during the invasion of southern France; she was broken up beginning in December 1945.

When the gunnery revolution began around 1900, the French navy was second only to the Royal Navy.[1] It considered itself more advanced in gunnery and capable of fighting at longer ranges. For example, the US naval attaché reported a practice at 5300 metres (5800 yards) (witnessed by the Minister of Marine, in which 13 per cent hits were made, well beyond what any other fleet could then have achieved. A French naval-gunnery text of 1899 included tabulated data for ranges up to 6000 metres (6560 yards).[2] It is not clear to what extent the French were motivated by the torpedo threat from battlehsips.

Then and later the French considered themselves very advanced in power control of heavy guns. Work on electric speed controls began in 1895, and the French developed the Williams-Janney controls used by most navies. After World War I the French navy became very interested in servo control (ie, remote-power control) of heavy guns. Thus the last two classes of battleships (*Dunkerque* and *Richelieu*) had servo control for train and elevation, although it may not have been entirely successful.[3] However, attempts at gyro-stabilisation apparently did not succeed until after World War II.[4]

Beginning in 1903 the standard rangefinder was the Ponthus-Therrode stadimeter.[5] The French navy rejected the early Barr & Stroud rangefinders, but by 1910 its standard free-standing type was the 9ft Barr & Stroud, and turrets were equipped with 4.5ft Barr & Stroud instruments which could be placed behind special ports. In 1914 Barr & Stroud delivered 15ft (FT 19) rangefinders. The following year the French navy bought 15ft triplex rangefinders (FT 25), which replaced 9ft instruments above the conning towers of battleships from the *Danton* class onwards. The French later considered these instruments too limiting on effective range, and they installed much longer-base ones in ships built after World War I.[6] After World War I the French navy switched to instruments made in France, the war having convinced the French government that it had to be independent in strategic materiel. Stereo rangefinders were adopted beginning in the late 1920s, and by the late 1930s remaining coincidence units were being replaced by stereo ones.

Battleships typically had double stereo units in their aloft directors (plus double units in their turrets: 12m (39ft) in the *Dunkerque* class, 13.5m (44.3ft) in the *Richelieu* class). Cruisers had single stereo units in their directors. At least in the battleships, one of the two stereo units could be used for what the French called *ecartometry*, the measurement of the vector between splash and target. This technique could also be used by cruisers with single rangefinders aloft.

Like others, the French tried a wide variety of transmitter-receiver systems, which became more crucial as gunnery became more sophisticated. Many were electric, because in 1900 France led the world in electrical engineering. Numerous French officers developed their own systems during the 1890s. The most successful was Eng's voltmeter system. Experiments with it began about 1890. In an attempt to standardise, it was adopted in 1898 (by which time it outnumbered the other systems then in use). Examples survived as late as 1908. An exotic alternative, the Germain hydraulic system, superseded it: the pressure indicated by a manometer moved a pointer around a dial. An 1898 report compared it – as installed on board the cruiser *Latouche-Treville* – with the Eng system on board the cruiser *Pothuau*.[7] Both seemed equivalent, and the Germain system was less than half as expensive. Germain's system was adopted, although Eng systems were ordered through 1900. Thus, after the tests, the battleship *Bouvet* was ordered fitted with the Eng system late in 1898; she retained it until a major refit in 1907. In 1900 about half the fleet had the Eng system, half the Germain. That year reports of battle practice showed that the Germain 'always operated with complete satisfaction' but the Eng was 'often deregulated or would simply cease to function'. By about 1906 there was considerable interest in replacing the Eng systems on the (approximately) forty ships that still had them.

The single hydraulic line of the Germain system supported one set of dials, so each had to display all the requisite data: firing orders, range (usually in hundreds of metres), and train, different data being sent in sequence. Precision in range was limited because range occupied only a fraction of the single dial. Maximum indicated range was 4500 metres (4920 yards). By 1906 the fleet wanted 8000 metres (8750 yards), and ultimately maximum range was 14,000 metres (15,310 yards).[8] Accuracy was 1 per cent, eg, 100 metres (109 yards) for a maximum range of 10,000 metres (10,940 yards).

The French dreadnoughts were all extensively rebuilt between the wars. *Paris* is shown here after her first major refit, October 1922–November 1923 at Brest. The massive tripod was intended to increase effective range, so her gun elevation was increased from twelve to twenty-three degrees (corresponding to a maximum range increased from 13,500 metres (14,760 yards) to 26,000 metres (28,400 yards). Inside the shelter atop the tripod was a director. Partly to compensate for the weight of the tripod, and also to improve seakeeping, armour was removed from her bow. The 13.4in gun fired a 432kg (952lb) shell at 783 metres (2568 feet)/second. Thus it was considerably heavier than contemporary US and British 12in shells.

By 1908 the Germain system was no longer well liked, mainly because the manometers of the after turrets showed readings different from those to which the dials were set because they were located on the other side of hot machinery spaces through which the single hydraulic line had to go (the fluid expanded as it passed through the machinery spaces). Attempts at insulation failed. The system was also disliked because it had to be bled daily before it could be pressurised for use. For the new *Danton*s, the French decided to discard the Germain system in favour of Barr & Stroud-type stepping motors. The French adopted the 'L-A System,' a version of the Barr & Stroud motor devised by Lieutenant de Vaisseau Lecomte and civil engineer Aubry of the Bourdon company. For the first time, observations or orders could be transmitted instantly. This L-A equipment, which appeared in 1910, was in effect a prerequisite for the new centralised fire-control systems.

Beginning in 1905 French ships had a 'central' control space below the armour deck under the conning tower, communicating with the bridge, conning tower, guns and magazines. It also handled damage control and was presumably a protected steering- and ship-control post. The fire-control element comprised simply telephones and signal lamps for conveying firing orders. An adjunct gunnery officer was stationed in the central.

Tests of new gunnery techniques began late in 1904. Half the Mediterranean Fleet tried Scott's continuous-aim method (see chapter 1); the other half tried a new French alternative. Similar experiments tested control by rangefinder (the *method telemetrique*) against an earlier concept of bombarding a 'beaten zone' (*zone battue*) through which the target was expected to pass. In the new method – like that the Royal Navy was then adopting – range was measured frequently and hand plotted, the officer at the transmitter allowing for dead time between measurement and firing. The difference between rangefinder and gun range would be estimated by spotting, after which the ship would use her rangefinders to set her sights.

Like the British, the French had a range clock and a rate solver (Dumaresq equivalent). The clock was the *indicateur continu* (also called *correcteur* and *pendule* [precision clock] *Lafrogne*) proposed in 1903 by Lieutenant de Vaisseau Lafrogne. He considered range-keeping one of the most absorbing tasks of the control officer; mechanising it would free him for other duties.[9] A sufficient number of units was ready by late 1906 for tests to be undertaken by the whole fleet. Lafrogne developed a complementary rate-solver (a *plateau*, or plate – a Dumaresq). For the tests, ships plotted rangefinder, clock and fall-of-shot ranges. The later standard Dumaresq equivalent was the *plateau* devised by Lieutenant de Vaisseau Lecomte, which appeared in 1909 (a 1911 version used a pointer to indicate rate).

The old cruiser *Pothuau* became gunnery training ship; new methods of fire control developed on board were being taught throughout the French fleet in 1908.[10] A typical range plot of this era shows a series of saw-tooth lines indicating clock ranges as they run and are reset (hence the jumps), plus a faired line of ranges derived from rangefinder readings and dots indicating the observed fall of shot.[11] The gunnery officer (*directeur*) and his assistant were stationed in the conning tower. Targets were verbally designated. Casemate guns had training marks on deck, presumably the remnants of an earlier director or concentration system. The French divided the action into a preparatory period, a probing (*recherche*) period, and a period of effective fire. During the initial period the gunnery officer (*officier de tir*) determined initial elements (range, bearing, deflection) and set up the clock and calculator. Sights were initially set at the rangefinder range, and translated into gun elevation by the officer's assistant, who used a table. This data was sent by voice pipe to the guns. Initially pointer and trainer were responsible for getting onto the target and for spotting, roughly every twenty seconds. The pointer set the sights. Control was entirely by voice pipe (which was ineffective against the noise of battle or for controlling multiple weapons), but some gun orders were given by individual electric lights (introduced in 1896). During the *recherche* phase the gunnery officer used salvo fire first to get line and then to get the range. By 1910 typical practice roughly matched the technique employed in the Royal Navy: firing salvoes to bracket the target, then shifting to rapid independent fire when it was straddled (a few years earlier there was no apparent interest in the final rapid-fire phase).

By 1910 it was standard to calibrate guns before practice, measuring their dispersion. Half the dispersion became the standard spotting step, the idea being to start short and then walk the salvoes back to the target. Using dispersion as the step was supposed to guarantee hits as soon as the target was crossed.

The French rejected American-style vertical spotting (see chapter 9, page 176) as well as target tracking to find and hold the rate (they had tried such plotting, but it did not work, particularly in the cramped quarters of their conning towers). American officers thought that, given their rejection of vertical spotting (which they considered inaccurate), the French had no real need for elevated fire-control platforms. They used a Dumaresq equivalent and a range clock, but without a plot their rate estimates could not be very good (the most charitable of the American officers thought the French were about two years behind the US Navy). An American officer reviewing French practices in 1911 remarked that they made sense only at the short ranges and under the very easy conditions (moored or very slow targets) at which the French exercised. At long range, where rangefinders were unreliable, the first shots might be as much as 1000 metres (1093 yards) from the target, and only bold corrections would bring them to the target quickly enough. Without tracking or plotting, initial rate estimates would hardly suffice.

In 1909 typical ranges for target practice were 5000 to 8000 metres (5470 to 8750 yards).[12] In 1911 American observers thought that 6000 metres (6560 yards) was still the expected battle range, although there were reports of practice firing at up to 12,000 metres (13,120 yards). They felt that the French had been influenced far too much by their earlier experience, having undergone little development while they lacked a dedicated naval gunnery school (between 1901 and 1909 naval artillery was controlled by the Colonial Troops, interested mainly in coastal defence). The school to be established at Toulon (by a decree of 1910) became the French naval-gunnery research and development centre.

The French navy practiced concentration by divisions, using one of two techniques. In the echelon method, each ship of the division fired at the target ten seconds after the previous ship, in order to find the range (while distinguishing splashes). This was

The French light cruiser *Lamotte-Picquet* displays the standard French interwar cruiser director at her masthead in this 27 February 1937 photo, taken at Shanghai. Just as the British AFCT was associated with a director control tower (DCT), the post-World War I French computer systems were associated with a DCT, in their case of cylindrical form. Like the British DCT, it had two levels, the upper for control and the lower for the pointer and trainer (at the front) and for rangefinding. Sights were apparently gyro-stabilised, and aim was automatically corrected for dead time. Unlike British DCTs, the French ones had an additional 3m (9.8ft) Zeiss stereo rangefinder at the front of the DCT, presumably mainly for spotting (*ecartometry* – measurement of the vector between splash and target). Thus the lower level showed a coincidence rangefinder at its rear, the short stereo unit near the front, and two large squared-off windows for the pointer and trainer. The first postwar cruisers (*Lamotte-Picquet* and *Duguay-Trouin* classes) had 4m (13ft) coincidence rangefinders, but in later classes the 5m (16.4ft) Model 1926 was standard, and in the 1930s ships were refitted with 8m (26ft) rangefinders. The level above had three windows, two together (probably, in analogy with the British, for the control officer and the rate officer), and one to the side (for the spotting officer), plus an armoured hatch, probably for an inclinometer. A prominent projection at the front of the cylinder carried periscopes for a cross-leveller and for the control officer. Early versions had the periscopes projecting upwards, but ships completed after about 1931 (*Algérie*, *Emile Bertin*, the 7600-ton cruisers) had two prominent periscopes projecting downwards, side by side. The 7600-tonners introduced a reshaped director with a prominent balancing vane running down its face. The main rangefinder (and the ones in A and X turrets) was an 8m (26ft) duplex stereo unit (two windows at each end). Apparently one of the two elements was used for ecartometry; there was no need for a separate spotting rangefinder. The DCT was introduced about 1927–8 (some cruisers were completed without anything at their mastheads). It equipped the rebuilt *Paris*-class battleships as well as new-construction cruisers. In addition to the 4m (13ft) rangefinder in her DCT, *Lamotte-Picquet* had another atop her bridge (visible here) and two more abeam the prominent ventilators in her waist. Note the prominent concentration dial near the foot of her foremast, and her small 152mm (6in) gunhouses, without rangefinders. She was lost in 1945, but her two sisters survived World War II. This class was armed with 155mm/6.1in/50 Model 1920 guns firing a 56.5kg (124.5lb) shell at 870 metres (2853 feet)/second. The later light cruisers had 6in (152mm) guns. *Emile Bertin* had the fifty-calibre Model 1930, derived from the 140mm (5.5in) Model 1929 aboard super-destroyers. It fired a 54kg (119lb) shell at 870 metres (2853 feet)/second at five to six rounds/gun/minute. The final French light cruisers (7600-ton class) and the *Richelieu* secondary battery used the same weapon.

considered complicated, so the preferred alternative was for each ship to fire at the corresponding ship in the enemy line in order to establish the difference between rangefinder and gun range, then shift targets and begin rapid fire, ships shifting back to ranging salvoes when they began to miss.[13]

Like the British, the French were well aware of the possibility that machinery could provide large-calibre weapons with something approaching continuous aim. Unlike the British, they used electric power. Guns up to at least 193mm (7.6in) calibre were typically fitted for power elevation, hence at least an approach to continuous aim, by 1913. By the outbreak of World War I similar continuous aim was being provided for 305mm (12in) guns, though not all ships had it.

In 1912 three artillery officers, Lieutenant de Vaisseau du Boucheron of the applied gunnery school

CHAPTER 13
The Italian Navy

The battleship *Roma* is shown nearly complete in 1942. As flagship of the Italian fleet, she was sunk by a German guided bomb en route to Malta in September 1943. Note her two masthead rangefinders, the lower unit intended for tactical ranging to support tactical plotting.

THE ROYAL ITALIAN NAVY developed its Dumaresq (equivalent) somewhat earlier than the Royal Navy.[1] With it they used a range clock and a range transmitter (the 'Ronca' system).[2] Ranges were corrected for known differences between rangefinder and gun performance using nomograms (graphs).

The Italian Navy bought the Barr & Stroud 15ft triplex rangefinder in 1916, a year after the French Navy, for all six of its dreadnoughts. By this time Italy had entered the war on the Allied side, and was receiving considerable British assistance. A US naval attaché visiting an Italian battleship in 1917 described what seems to have been a Dreyer turret table (range-only), using Dreyer's rotating-grid technique to derive the range rate and insert it into a clock. The Italians may also have had an equivalent to the Dreyer corrector.[3] In October 1917 BuOrd offered the Italian naval attaché access to the Ford Mk I and II Range-keepers, as they had been shown to the British and asked for by the Russians. It does not seem that they made much impression on later Italian developments.[4]

In September and October 1919 several US officers visited the Italian battleship *Conte di Cavour* when she visited the United States.[5] The ship's gunnery officer described his system as modelled closely on the British (she did not yet have director control), with the gunnery officer stationed in the conning tower. He maintained his own plot there, from which he derived the range rate. The range rate was also maintained in the plotting room, using a range clock and a plotting board (projection-component method, presumably as in a Dreyer Table). Should clock and plot agree, the gunnery officer used this range rate. If they disagreed, he used his own estimate. Although there were two spotters aloft, the gunnery officer placed little reliance on them, preferring what he could see through his own

At Constantinople in August 1919, the battleship *Andrea Doria* shows a triplex rangefinder atop her bridge and a small spotting top forward. By this time she and other Italian battleships had been equipped with French Le Prieur tables, broadly equivalent to Dreyer Tables, and they were receiving Girardelli data transmitters. Within a few years the Royal Italian navy would abandon such equipment in favour of the British Barr & Stroud system.

This post-World War I photograph of the cruiser *Kirov*, the first new major combatant built by the Soviets, shows the Italian-inspired triple-rangefinder KDP-6 director at her foretop. She was armed with a new high-velocity 180mm/7in/57 gun conceived in 1925 to fire a 100kg (220lb)shell at 1000 metres/3280 feet/second (it actually fired a 97.5kg/214.5lb shell at 920 metres/3018 feet/second). A single mount (MK-1-180) armed the ex-Tsarist cruiser *Krasni Kavkaz*, completed in 1932; the *Kirov*s had the triple MK-3-180. Note that MK meant Morskoi Kanone, not Mark as in Western navies. This was the most powerful weapon to arm a Soviet-built ship, plans for ships armed with guns of 220 to 406mm (8.6 to 15.9in) calibre being either stopped by World War II or by Stalin's death in 1953. Maximum elevation was an unusual – for cruisers of the 1930s – forty-eight degrees. Claimed rate of fire was 2 to 5.5 rounds/gun/minute. The mount and gun were apparently unsatisfactory, because the next class of ships, the *Chapaev*s, had 152mm (6in)/57 guns; this calibre was also used for the redesigned version of the *Chapaev* class, the *Sverdlov*s built postwar.

fire, so they began to develop techniques of their own. They soon concluded that no more than three ships should fire together at the same target, for fear of confusing their splashes. Now there was a different reason to try concentration: two or three pre-dreadnoughts might equate to a single dreadnought if they could coordinate their fire. The denser the salvoes, the better the fire control at long range. The objective was completely to centralise control of the fire of the group. By 1914 the Black Sea Fleet had a brigade of three 304mm (12in) pre-dreadnoughts, *Evstafii*, *Ioann Zlatoust* and *Panteleimon* trained in concentration tactics, the middle ship of the group acting as master ship and transmitting range and deflection by radio to the other two ships (special antennas were spread on bamboo supports on each side of the master ship, so that hits on one side would not disrupt them).[11]

Concentration tactics were used in a battle between this Black Sea Fleet brigade and the German battle-cruiser *Goeben* off Cape Sarych on 18 November 1914.

It revealed some weaknesses. *Zlatoust* was master ship, but could not see the target due to mist, so *Evstafii*, which could see it, did not open fire. When *Zlatoust* did transmit an opening range, it differed radically from that which *Evstafii* had measured (sixty cables/12,000 yards as opposed to forty cables/8000 yards); *Evstafii* was ordered to open fire using her own data. She hit on the first salvo, and nearly caused a magazine explosion. *Goeben* made several hits. The other Russian ships either used inaccurate ranges from *Zlatoust* or never fired. Russian gunnery impressed the Germans; the salvoes were so well concentrated that the German commander thought he was under the fire of five battleships, and he fled. This was pretty good for a single pre-dreadnought fighting a battlecruiser.

Spotting and bracketing entailed considerable calculation, some of which was mechanised in the next Geisler system (Model 1911, tested 1910 on board the battleship *Petr Velikiy*). Like the British, the Russians concluded that the central station (*zentralnii artilleriiskii post*, ZAP) had to be protected and placed below decks (equipment could not be accommodated in the conning tower). The ZAP had a range clock which automatically transmitted range to the gun sights. It is not clear to what extent range data were transmitted automatically or by telephone from rangefinder to central station. In the dreadnoughts both range and target bearing were transmitted by telephone and corrected in the ZAP (eg, deflection corrected for drift). As in the earlier system, data from different rangefinders were averaged in the central station to produce a single range (which might be updated by range rate) for onward transmission to the guns. At each of the guns was a cam translating range into elevation and a lever which corrected elevation for individual gun wear. The central station also corrected both range and deflection for time of flight, which for a 304mm (12in)/52 could be as much as eighty seconds. In the dreadnoughts, the translation into elevation was apparently done in the central station, using ballistic tables (in a pre-dreadnought, it would have been simpler to transmit the same range to guns of different calibres). Clock settings could be corrected by spotting. Initially range rate was either estimated by eye or measured by entering successive ranges into a range clock. This system introduced follow-the-pointer operation in both elevation and bearing.

By 1914 ships plotted enemy range and bearing before opening fire to estimate enemy course and speed and thus range rates. It was assumed that rangefinders would function badly after fire was opened. In effect the Russians were betting that range rate was constant and that it would remain constant long enough to make it useful.[12] Under these circumstances there was little point in armouring long-base rangefinders. The *Gangut*s were completed with a 5m (16.4ft) Zeiss rangefinder atop each conning tower, plus three smaller units (one Barr & Stroud 4.5ft and two Krylov 'Model 1911' stadimeters). When the ships were laid up at Helsingfors (Helsinki) during the winter of 1915–16, however, the 5m (16.4ft) rangefinders were moved to the end turrets, and 5.5m (18ft) Barr & Stroud rangefinders were placed in the second and third turrets. Installation of rangefinders in the turrets seems to have been intended to provide local control. The Russians also wanted to increase the number of rangefinders so that errors would be easier to average out. The Russians then ordered 2.7m (9ft) Pollen-Cooke rangefinders, only three of which were delivered (the Admiralty blocked further orders). One was fitted to *Gangut*. Of the Black Sea ships, only the last, *Volya*, ever had turret rangefinders (one 5.5m (18ft) Barr & Stroud on each turret).[13]

The Geisler system was introduced in the last two Russian pre-dreadnoughts, the *Imperator Pavel I* class. Their gunnery-control spaces were under their conning towers, an armoured tube connecting the two. These ships also had lattice masts, presumably an attempt to avoid the loss of aloft fire-control positions experienced during the Russo-Japanese War; such masts were not used in any other Russian ships however.[14] The 1911 Geisler system was widely used in the Russian navy, and it was installed in the first generation of Soviet ships as well; it survived in Soviet service into the 1960s.[15]

The Geisler system was limited, and it could not work from a manoeuvering ship. Around 1910 a Russian firm, Erikson of St Petersburg, managed by retired Colonel Ia A Perepelkin, began development of a centralised fire-control system employing calculators and centralised control. This Model 1912 system would have equipped the fourth (unfinished) Black Sea battleship, *Imperator Nikolai I*.[16] It appears that enemy course and speed would have been calculated graphically, based on observed range and bearing, the calculator taking into account own-ship course and speed as well as corrections for wind, drift, sight elevation (dip) and parallax. The artillery officer would fire a salvo using a single key; the junior artillery officer in the foretop would spot splashes. According to the British, in the autumn of 1916 the Baltic dreadnoughts (*Gangut*s) were equipped with an Erikson director system for their main batteries.[17] Russian writers do not confirm this.

In 1911 the Naval Technical Committee, which was in charge of ship design, decided to use the Erikson

PROPELLANTS, GUNS, SHELLS AND ARMOUR

The Imperial German navy planned to fight at short ranges, so it chose high-velocity guns firing relatively light shells. These 12in/50s are on board the battleship *Ostfriesland*, under the US flag having been turned over at the end of the war. A US officer who visited a German 12in turret (on board SMS *Kaiser*) in July 1913 was struck by how roomy it was, without any bulkhead between the guns or any separate officer's booth at its after end. The back of the turret was left clear so that the guns could be hand-rammed through openings in the rear. The officer stood between the guns, near the front of the turret. The 4m (13ft) turret rangefinder used large roof hoods (visible here on the turret roof), which the US observer felt exposed it unnecessarily to blast. Sight holes (for pointers' secondary telescopes) were cut in the face of the turret and, in some cases, the usual turret hood – for the officer – was replaced by another opening in the turret face. Unlike US turrets, the German turret used a horizontal car for its shell, making ramming into the breech simple and rapid; there was no danger of the shell going adrift if the ship rolled heavily (later ships would receive tube hoists, in which the shells ascended vertically). German officers said that they preferred hydraulic to electric operation because it was easier to detect and fix problems. There was no separate handling room, the powder hoist extending directly down into the magazine. After the Dogger Bank action, some safety precautions were taken. The ready service stowage of shells and cartridges (six shots) previously kept in the turret was eliminated. Double-flap doors were placed at the top and bottom of the cartridge hoist to prevent fire from passing down into the magazine. A US officer who visited the battlecruiser *Moltke* in January 1915 was told that the system was designed for particularly rapid fire; he witnessed a drill that would provide better than four aimed shots per gun per minute (another paper gave a firing interval of twelve seconds). Below the weather deck can be seen the ship's 5.9in secondary guns, which the Germans planned to use to supplement the heavy guns. This practice helped convince the British that the Germans would not fight at long range. German fleet manuals expressing tactical doctrine carried much the same message. The crow's nest visible on the mainmast was a spotting position.

became possible to make cartridges for guns larger than small arms: a new class of quick-firing (QF) naval guns came into existence. Where a screw-breech, 4.7in gun could fire about a round a minute, a QF gun of similar calibre might fire five.[3]

The Germans were unique in using QF methods for all calibres. No single case could be massive enough for a heavy gun, so the Germans used a full brass case only for propellant near the breech, where the gas seal was most important. The other charges were cased in thin zinc, which was destroyed when the gun fired (many imagined, incorrectly, that the rest of the charges were in bags, protected by cases until they were loaded). The

Germans credited their fully cased ammunition with saving them from explosions when their battlecruisers took turret hits at Dogger Bank and Jutland.

From about 1890, improved breech mechanisms made it possible to fire bag guns more rapidly, and thus to rival QF weapons. The US Navy, for example, used bagged propellants for the high-velocity 5in/51 it used for battleship secondary batteries from about 1910 onwards. Similarly, it used bags for the 6in/53 of the *Omaha*-class light cruisers. However, bagged ammunition could not be used in an automated loading cycle. The US Navy adopted cased ammunition in the 6in/47 of the late 1930s and then in the automatic 8in/55 of the *Newport News*-class heavy cruisers. One barrier to adopting cases for the propellants for large calibres was that they were too heavy for manhandling.

Bofors found a neat solution in the 5.9in gun it developed in the late 1930s for the Dutch *de Zeven Provincien* class and the Swedish *Tre Kronor* class. Case and shell were manhandled separately, but before approaching the gun the shell was crimped into the cartridge case to form a single unit that could be easily handled (by machinery).

Rapid fire would have been pointless had the target quickly been obscured by the kind of smoke that black or brown powder produced. Thus rapid-firing guns were associated with a new smokeless (actually reduced-smoke) powder. The French were the first to produce smokeless powder. Typically they combined two explosives, nitroglycerine and nitrocellulose (guncotton) with a plasticiser and a stabiliser, hence were called double-based. Because these powders were altogether unarmoured. Some ships sank because they were not stable enough, and a few hits at the ends brought unarmoured parts close enough to the waterline that the hits shattered them. Many of the Russian ships had not been stripped properly before battle, and they burned. DNO and his successors were impressed that, as in the Boer War, modern high explosives had a kind of chemical effect, staining everything near the hit yellow. Personnel near the explosion who had not been injured directly complained of stupor, giddiness, headache and loss of memory, even twenty-four hours later.[10]

Ideally an AP shell would survive penetration and burst inside, using a delayed-action fuse activated by the shock of hitting armour. The same shock could not be allowed to set off whatever explosive filled the shell. There was clearly a fine line between a sufficiently powerful detonator and one so sensitive that it would prematurely detonate an AP shell. The British Grand Fleet battle orders drew fine distinctions between which kinds of shells ships should fire, depending on the range at which they would be firing.[11] During World War I the Germans differentiated between AP for shorter ranges (where they could penetrate) and HE for longer ones, but its wartime success with AP at long range impressed other navies, such as the French. The US Navy rejected the HE argument altogether, limiting itself to AP shells (apparently some had gas-producing filling instead of pure explosives). Later it adopted HE and High Capacity (HC) shells specifically for shore bombardment, which became an important function during the interwar period.

TNT, which the Germans adopted about 1902, was the first truly insensitive explosive.[12] Setting off such an explosive required a shock stronger than that which the usual black-powder fuse could provide. The solution, devised by the Germans by 1911, was a two-stage fuse. The shock of hitting set off black powder, which in turn set off a secondary, booster or gaine charge. This two-stage process also made it possible to design a fuse with a delay, so that the shell might explode well inboard, within a ship's vitals. The two-stage fuse required precision production processes. The Germans used black powder to detonate a picric-acid booster. A long twisting chamber from primary to booster was intended to impose the delay, although it did not always work as intended. The Germans adopted their new fuse, and a TNT filler, in 1911. The Austro-Hungarians followed suit (Krupp and Skoda, the two manufacturers involved, had a close relationship).

British shells failed to perform at Jutland, typically either exploding on impact or breaking up (only two 15in APC shells penetrated and exploded behind armour as intended). Two types of shell were then in use, APC filled with Lyddite (using a delayed-action fuse) and powder-filled CPC (Capped [Armour] Piercing Common) using an instantaneous fuse. Apparently there were three problems: Lyddite was (as had originally been thought) too sensitive; shell steel was too brittle, particularly given caps designed for impact at right angles rather than obliquely; and the fuses were poorly designed.[13] Initially the British thought the fuses were the only problem, and tried to copy German fuses; but they soon realised that much more remained to be worked on.

The German cap was copied, and shell steel subjected to further hardening. Shellite, a new explosive filling combining dinotrophenol (DNP) and Lyddite, was adopted.[14] By 1918 the British had a new delayed-action fuse using a pellet of compressed powder as its gaine. Shell Committee tests suggested that it gave a reasonably reliable thirty-five-foot delay. In 1919, when the Shell Committee delivered its final report, the US Navy did not yet have a reliable delay fuse for AP shells, and the British were trying to make their pellet fuse reliable. The US Navy was considering adopting the German fuse. In the early 1920s the Germans offered the US Navy their wartime ordnance developments, and BuOrd bought only the fuse.

After World War I both the Royal Navy and the US Navy inserted a third stage between delay and booster, weakening the primary charge. The Imperial Japanese navy designed its later fuses (for Type 88 and Type 91 shells, introduced in 1928 and in 1931) for unusually long delays for the underwater hits desired (the shock of entering the water would otherwise set the fuse

Beginning about 1933, the US Navy's Bureau of Ordnance became interested in heavier shells that might be more effective at long range. In September 1933 it circulated a study showing that greater weight (1500lb versus 1400lb for 14in guns, 2250 versus 2100 for 16in) would cut range, but that increased weight plus better streamlining (longer ogives) would gain it back. Thus maximum 14in/50 range would increase from 35,700 to 36,800 yards. A 14in/45 using the new shell would penetrate 13.5 in side armour (as used on US battleships) anywhere inside 21,000 yards (rather than the previous 14,500 yards), and the usual 3in deck would be penetrated anywhere outside 20,000 yards, so a typical US battleship would have no effective protection at all. By 1939 the bureau was interested in much heavier shells. Existing 14in hoists could not accommodate anything beyond 1500lb, but the new battleships then under construction could easily be modified to fire 2700lb shells (the most the *Maryland* class could handle was a 2240lb shell). A 7 June 1939 bureau memorandum to the General Board observed that the 2700lb shell would approach the performance of a 3200lb 18in shell (which, it turned out, was roughly what the Japanese *Yamato* class fired). Forty such shells had already been fired, demonstrating that dispersion was somewhat worse than for the lighter gun (at sixteen-degree elevation, 0.25 per cent rather than 0.16 per cent of range, but at forty-degree elevation – extreme range – 0.27 rather than 0.30 per cent, – ie, an improvement). Figures were also given for super-heavy 14in shells (1800 and 2000lbs), but they could not be accommodated on board existing ships. The heavier shells were not expected to affect gun lifetime. Maximum range, after firing 100 full-charge rounds, would be 38,800 yards for a 16in/50 rather than 43,000 yards with the 2240lb shell (at maximum elevation, forty-five degrees, range for the 16in/50 would be 41,700 yards, compared to 45,900 with the lighter shell and 43,900 yards for the 18in/48 firing a 3200lb shell). On the other hand, the target would lose about 6000 yards of immune zone. With the heavier shell the 16in/50 would penetrate 14in side armour at 32,000 yards, compared to 31,300 for the lighter shell and 35,600 for the 18in gun. It would penetrate a 6.5in deck anywhere beyond 29,300 yards, compared to 33,200 yards for the lighter shell and 29,700 for the 18in – ie, it would outperform the 18in for deck penetration. It showed similar superiority in deck penetration against thicker decks. These figures were for a ninety-degree target angle; note that the Royal Navy used a more conservative seventy-degree angle, at which side penetration would be reduced. Given such performance, on 26 June 1939 the General Board formally recommended adopting the heavy shell. Two days later the Secretary of the Navy approved the change. The heavy 8in (204mm) shell (335lb versus 260lb) was approved on 25 January 1941; in its case length was held constant by using a somewhat blunter nose. Ships like *Nevada* spent most of their wartime careers delivering fire support rather than dealing with enemy capital ships. Delivery of special bombardment (high-capacity) shells began in October 1942; weights were 975lb for 12in (*Alaska* class only), 1275lb for 14in, and 1900lb for 16in, ie, lighter than AP. The new 12in and 14in shells did not fit the hoists of the *Arkansas* and *New York* classes, but they were carried by all later ships. Plans at the time were for ships to carry 85 per cent AP and 15 per cent HC shells. Later the proportion changed, so that at Surigao Strait there was real concern that ships might not have enough AP on board to deal with the approaching Japanese battleships. A triple 16in/45 turret (No 3) is shown aboard the battleship *Alabama* at Puget Sound Navy Yard on 25 February 1945, for the ship's post-refit inclining experiment. Note the sights on the side of the turret (separate ones for pointer and trainer, with similar ones on the other side of the turret), and the port end of its long-base rangefinder.

working). Postwar US and British analysts considered that a mistake, because it further reduced the chance that AP shells would explode after hitting unarmoured ships. This feature may help explain the survival of the US destroyers and destroyer-escorts under battleship fire off Samar in October 1944.

Guns

Overall, the gun designer chooses between initial high velocity (a long gun) and shell weight. Long guns sometimes drooped, or showed undue dispersion (ie, the shells fell too far from the point at which they were aimed). The British 12in/50 in particular was credited with too short a service life and with excessive dispersion (due to its lack of rigidity).[15] The British therefore chose larger-calibre guns firing heavier shells: the 13.5in/45 and then the 15in/42 and the 16in/45. They were inherently more rigid, and at longer ranges they retained their velocity. The US Navy followed the British towards larger calibres, but it also liked long guns. It followed its 12in/50 with a 14in/45 and then with a 14in/50, before it adopted a 16in/45 (ships cancelled under the Washington Treaty would have had 16in/50s). After World War I the US Navy developed what it thought was a unique capability to fight at extreme ranges, where shells would hit mainly deck armour. The heavier the shell, the more velocity it would retain when it hit. The US Navy therefore chose moderate muzzle velocities and what it called super-heavy shells (2700lb rather than the previous 2100lb for 16in guns). US calculations showed that a 16in gun firing a super-heavy shell would be as effective as a more conventional 18in gun.[16] In effect Japanese adoption of the 18.1in/45 in the *Yamato* class reflected

British thinking – that the way to achieve better performance at long range was to make a shell heavier. As in World War I, the Germans opted for high velocity in their 15in/48.6 (with a considerably lighter projectile than that of the British 15in/42) and 11in/51.25 (sometimes given as 11in/54). The French chose high velocity in their 13in/50 but lower velocity and a heavy shell in their 15in/45. The Italians had a high-velocity 15in/50 firing a heavy shell. The Soviets planned a 16in/50 for the battleships they never completed.

Shell weight is proportional to shell volume, ie, to the cube of calibre. As a rule of thumb, a typical shell might be half the cube of calibre, in pounds. In that case a typical 10in shell would weigh 500 lb.

Nominal weights for larger calibres, compared to some actual AP shells, are given in the table below:[17]

The longer the curved shell nose, the better the aero-

	6in	8in	11in	12in	13.5in	14in	15in	16in	18in
Nominal weight in lbs	108	192	665	864	1230	1372	1688	2058	2908
Actual AP shells									
US Navy (heavy series)	105	260	—	870	—	1400	—	2100	2900
Royal Navy	112*	256	—	850	1350	1590*	1920	2048	—
Germany (WW I)				626	915				
Germany (WWII)	100	269	728	915**	—	—	1764	2271	—
Japan (WW II)	123	277	—	883	—	1485	—	2249	3219
France (WWII)	119	295	—	952	1268***	—	1949	—	—
Italy (WW II)	110	276	—	1157****	—	—	1951	—	—

* World War II shells.

** Design, not built.

*** This is the World War I 13.4in gun with an M1924 shell. The 13in gun of 1930 fired a 1235 lb shell.

**** This is the World War II 12.6in gun bored out from earlier 12in guns that fired a 997 lb shell.

dynamics. Usually the nose is a thin metal windshield carrying very little of the mass of the shell. Nose curvature is typically measured in shell diameters, expressed as crh, calibre radius head. The Royal Navy used a mixture of 2crh and 4crh shells before World War I, having adopted 4crh in 1908 to reduce drag. By World War II the French and Japanese navies were reducing shell drag by tapering the after end of the shells into a 'boat tail'. Some modern shells smooth their air flow by releasing gas from the after end ('base bleed').

A gun showed random dispersion, measured in mils, thousandths of a radian (in effect, thousandths of range). Dispersion is due partly to the way that the gun flexes as it fires, an effect sometimes called 'jump'. Vibration at the muzzle (probably a particular factor in a wire-wound weapon) imparts a sideways kick. Shells rarely emerge quite centred from the gun, because they do not fit tightly in the barrel. A shell was sealed in the barrel by a driving ring near its base (using the ring rather than the whole of the shell reduced friction losses in the barrel). Thus the spin, which is intended to stabilise the shell and to keep it from popping upright once it emerges from the barrel under the effects of pressure and gravity, has complex effects. The shell behaves like a slightly off-centre top once it leaves the barrel, wobbling (precessing) a few degrees (generally less than ten degrees). Moving slightly sideways, it presents a larger area to the air slowing it down. To make matters more complicated, the yaw angle itself increases and decreases as the shell flies. If the shell is spinning at the correct rate for its onward speed, the wobbling dies down completely, just as a top eventually 'goes to sleep' or stops wobbling. If not, the wobbling may increase. The shell is slowing down as it flies. At some point it may be over-spinning, hence yawing quite noticeably. That may happen at the peak of its trajectory, in which case yaw may cause much increased air resistance.[18]

In addition to its other effects, the spin on the shell generates a sideways aerodynamic force that causes it to drift to one side. A fire-control system calculating deflection had to compensate for the drift rate over the expected time of flight of the shell. Given yaw and drift, two shells fired at about the same moment by two adjacent guns could collide in mid-air; they were said to 'kiss'. That was aside from interference between the shock waves of adjacent guns, which was a particular problem in triple and quadruple turrets.

Notes

Please refer to the abbreviations listed on page 6.

1 The Gunnery Problem

1 Figures from the 1910 official British range tables, ADM 186/181.

2 From *Range Tables 1918*; muzzle velocity was 2,525 feet/second. The higher-velocity 12in/50 (2,825 feet/second) reached this point at 17,100 yards. A 15in/42 reached the same point at 18,200 yards. By 1918 the Grand Fleet was firing at 24,000 yards.

3 John Brooks explains the consequences of the irregularity of the roll in 'Percy Scott and the Director', *Warship 1996*, p 154.

4 At rest a pendulum tries to point to the centre of the earth. Otherwise it swings around the direction to the centre of the earth; at the very least it defines a vertical direction without reference to the sea or to a ship swinging back and forth overhead. The gunner can judge from the swing of the pendulum where the true vertical lies, and the moment at which the ship is vertical. The guns are fired at just that moment. Unfortunately this approach requires precise judgement and very good timing, and small inherent delays ensured that the gun was never quite at the end of the roll when it was fired. It is not clear to what extent such pendulums were adopted by various navies. Admiral F C Dreyer's manuscript historical notes on British fire-control history begin in 1829 with a primitive method of director fire in which the moment to fire was determined by a 'marine theodolite' using a pendulum. The technique was invented (and published) by William Kennish, Carpenter, RN, and tested on board HMS *Hussar* at Bermuda. The Dreyer manuscript was provided by Professor Jon Tetsuro Sumida, and a copy of the Kennish paper is in the US Navy Library (Washington Navy Yard). In the first edition of his *Treatise on Naval Gunnery* (London: John Murray, 1820), pp 218–19, General Sir Howard Douglas attributed to Captain Sir Philip Brooke of HMS *Shannon* (a major Napoleonic-war gunnery innovator, known for his dramatic 1813 victory over USS *Chesapeake*) the idea of using a pendulum to determine the inclination of a ship (so as to correct gun elevation), but not her position in the roll (presumably the momentary position of the pendulum could not be communicated to all the guns quickly enough). Brooke's idea was generally adopted by the Royal Navy. According to Sir Howard Douglas, *Treatise on Naval Gunnery* (1855 [fourth edition] reprinted by Conway Maritime Press, London, 1982), p 387, the French went a step further and used the pendulum to indicate the moment when the gun was horizontal. According to Douglas, they considered the pendulum a vital discovery. The French used a reflector to compare the indication of the pendulum with the real horizon; this combination was called *L'Horizon Ballistique*. The US 1933 naval gunnery manual cited the French device (as described by Douglas) as a forerunner of director firing (to cancel out roll). Douglas was the son of Captain Sir Charles Douglas RN, whose innovations contributed considerably to Admiral Rodney's 1782 victory in the Battle of the Saintes (see Peter Padfield, *Guns at Sea* [London: Hugh Evelyn, 1973], p 111).

5 Gyros work because angular momentum is conserved. The angular momentum of the gyro is a vector along its spinning axis. When it is pushed from one side, it tilts at right angles to the push (precesses) so that the vector total does not change. It is this that makes bicycles stable as long as their wheels are spinning. A gyro on the earth's surface aligns itself with the earth's axis of rotation, and thus becomes a reliable compass (motion about its horizontal axis is suppressed by a weight). A shipboard compass wanders because of the motion of the ship; it does not instantly cancel a ship's turn. Brooks, DGBJ, p 29, reports that the Anschutz used by the Royal Navy on the eve of World War I wandered by five to ten degrees in anything but smooth weather, and by two degrees when a fast (twenty-five-knot) ship made a four-point (forty-five degree) turn. The 1912 version, redesigned to eliminate large errors in turns, is described in detail beginning on page 69 of the 1912 report of the Torpedo School (which was responsible for electrical equipment, including fire-control equipment). (Report courtesy of Dr Nicholas Lambert.) The 1913 Torpedo School report (NARA II) describes the replacement Sperry gyro-compass, two of which were installed for trials in the battleship *St Vincent* and in the submarine *E1*. During comparative trials at the Royal Naval College, Greenwich on a violently rolling and pitching platform, the Sperry compass kept its direction (north-south) within one degree, whereas the average error of the Anschutz was 3.7 degrees, with an error rate worse than one degree in five minutes. Both yawed rapidly within about one degree, although the platform was not yawing at all.

6 According to Professor Jon Tetsuro Sumida, IDNS, p 49, the first move in this direction was the Vickers BVIII mounting for the battleships of the 1904–5 programme.

7 DNO notes for his successor, July 1907, p 15. The elevating mechanism was being tested on HMS *Drudge*. Document provided courtesy of Dr Nicholas Lambert.

8 Brooks, DGBJ, p 46, quoting Captain A Craig, 'Rough Weather Testing, HMS *Orion*,' 15 November 1912.

9 Both Harding's book and the 1904 report are in NHB.

10 Both British and German dreadnoughts carried significant numbers of underwater torpedo tubes. Torpedoes seemed so promising that the US and Imperial Russian navies (and possibly others) considered building 'torpedo battleships'. Both backed away because these weapons had inherent limitations. The US Navy, however, doubled the number of underwater tubes in the *New York* and *Nevada* to four, despite the cost in ship size and in vulnerability to underwater hits. US-battleship torpedo tubes were removed when ships were modernised (they were also removed from the unmodernised 'Big Five'). By 1938, of British battleships, the two *Nelsons*, the five unmodernised *R*-class ships, and two unmodernised *Queen Elizabeth*s (*Barham* and *Malaya*) all had torpedo tubes, as did the battlecruisers *Hood* and *Repulse* (*Renown* apparently had hers removed on reconstruction). (Data from the April 1938 Armament List, CB 1773(38), PRO.) The World War II German navy mounted torpedoes on some battleships primarily to sink merchant ships, the theory being that guns were an inefficient way so to do.

11 According to Professor Jon Tetsuro Sumida, IDNS p 50, the first Mediterranean experiments were conducted in 1898 by Fisher's predecessor, Admiral Sir John Hopkins, leading to the salvo-firing concept. However, a lengthy discussion of improved gunnery training in the 1899 edition of *Principal Questions Decided by DNO* (ADM 256/35, pp 102–39) does not mention range at all.

12 See, for example, Sir Alfred Phillips Ryarder, *Methods of Ascertaining the Distance from Ships at Sea* (Portsea, W Woodward, 1854). (Reference courtesy of Stephen McLaughlin.) The Ryarder technique is mentioned in the 1885 official British gunnery handbook (ADM 186/896) as though it were standard. A contemporary source is General Sir Howard Douglas, *A Treatise on Naval Gunnery 1855* (reprint of fourth edition by Conway Maritime Press, 1982, pp 377–85). Douglas mentions one method, using the height of a ship's own mast, as the one already in use at HMS *Excellent*. His first edition (1820, pp 211–16) described rangefinding based on the known heights of a target's masts (he included a helpful table of standard French masts, observing that many American masts had the same dimensions).

13 According to his memoirs, Fiske became interested in rangefinding because he knew Lieutenant Zalinski, the inventor of the 'dynamite gun', which propelled projectiles using compressed air (the US Navy mounted these weapons on board the 'dynamite cruiser' USS *Vesuvius*). It had a steep trajectory, hence needed accurate range data. This weapon was abandoned because of its limited range. Later Fiske developed a stadimeter, which the US Navy used for fire control during the Spanish-American War.

14 The issue was raised by the Ordnance Committee. HMS *Excellent* advised DNO (Captain John Fisher) that the best rangefinder was a QF gun, but Fisher persisted. See Professor Jon Tetsuro Sumida, IDNS p 72, and DNOQ for 1889–91. The report on alternative rangefinders, is, surprisingly, found in the 1913 *Technical History* comparing the Pollen and Dreyer fire-control systems (in NHB). That was probably because Dreyer saw it as proof that, as he persistently claimed, Pollen's plotting had been tested more than a decade earlier by the rangefinder committee, as part of the Watkin system.

15 Barr & Stroud designations consisted of a function letter (eg, F for Range Finder), a sequence letter within that function (A for a 4.5ft instrument), and a sequence number. According to the company history, Michael Moss and Iain Russell: *Range and Vision: The First Hundred Years of Barr & Stroud* (Edinburgh: Mainstream, 1988), p 26, the first order, for six rangefinders, was placed late in 1892, the first being delivered in the summer of 1893 and installed on board HMS *Blenheim*. The sixth was delivered to the first foreign customer, Japan, for their new cruiser *Yoshino*, which used it successfully during the Sino-Japanese War (1894–95). Production instruments were modified, their length reduced to 4.5ft. By January 1896, sixteen had been delivered to the Admiralty, and thirty-six more were on order, most being the improved FA2. The Admiralty ordered a further fifty in November 1897, and another 100 in June 1899 (so that every major ship could have two, one forward and one aft). Under an agreement with the major British warship export builder, Armstrong, Mitchell, Barr & Stroud marketed to foreign fleets instruments soon being installed in the cruisers *Buenos Aires* (Argentina) and *Blanco Encalada* (Chile). By November 1897, twenty-seven had been delivered to various customers. By March 1898, 150 had been ordered by various customers. Japan was the main export customer, with forty-seven on order or delivered as of February 1901. Spain bought eleven when going to war with the United States in 1898. Argentina and Chile bought large numbers at about this time. The Admiralty ordered a 6ft instrument in March 1901 to be accurate to within 3 per cent at 6000 yards, but trials showed that the copper tube was not rigid enough (see Moss and Russell… p 44). With magnification increased from 20X to 24X in the FA3 version (1903), the 4.5ft instrument met the 6000-yard requirement. The Admiralty, which had ordered only eleven FA2s between June 1899 and September 1903, ordered numerous FA3s and had existing FA2s converted. The successor 9ft series were designated FQ, and the 15ft type were FT.

16 According to their 1917 handbook on German ordnance, the British thought the Germans had abandoned stereoscopic rangefinders before the war. That was probably disinformation; German policy was apparently to manufacture one type for internal consumption while marketing an alternative abroad. Both German manufacturers, Goerz and Zeiss, made both stereo and coincidence instruments. Pre-war US naval intelligence reports show that foreigners were told that the coincidence type equipped the Imperial Japanese navy. This was much like the policy under which the German navy used MAN four-stroke diesels (which worked) while the firm sold less reliable two-stroke diesels abroad.

17 The range error, in yards, is given by the product of the angular limit and the square of the range, divided by baseline length multiplied by magnification and by 206,265 (a factor to convert arc seconds to radians). Here range and baseline are in yards, and power is in diameters. The angular limit is typically twelve seconds of arc, although it is sometimes given as fifteen arc-seconds. Thus to maintain the same performance at greater range the designer can either extend baseline or increase magnification. Because the error rises as the square of the range, it takes four times the improvement to maintain accuracy at twice the range. The formula, courtesy of Chris Carlson, is taken from the ASNE Journal for February 1920 (Vol XXXII, No 1), pp 1–37; it is also given in the September 1950 edition of the US Navy's NAVPERS 16116-B, *Naval Ordnance and Gunnery*, pp 333–4.

18 After 1906 trials, the 9ft Barr & Stroud was selected over a 10ft rangefinder by Cooke of York. The Admiralty hoped that Cooke would become a viable alternative to Barr & Stroud, but its prices were never low enough. The threat of competition may have forced Barr & Stroud's prices down. Cooke was later bought by Pollen; the Argo Pollen-Cooke rangefinder was of a different design, with claimed advantages at low light levels.

19 Admiralty Technical History, *Fire Control in HM Ships* (Pt 23), dated December 1919, p 33 (ADM 275/19). Larger errors were reported in combat (eg, 1500 yards at 16,000 yards at the Falklands), but they were estimates, whereas the March 1917 tests were against a target at known range. These errors were still unexplained as of late 1919; refraction was specifically rejected. In these tests, 9ft masthead rangefinders outperformed longer-base ones in turrets.

20 Brooks, DGBJ, p 51.

21 According to the Barr & Stroud company history (p 79), DNO Jellicoe encouraged development of a 15ft rangefinder to extend ranges to 20,000 yards. Several navies ordered 15ft trial models (Model FR) between 1907 and 1913, after which the FT24 version was introduced as

invented independently in 1901 by D P Thompson of GE and by C A A Mickalke of Siemens-Halske (from the BuOrd wartime fire-control history manuscript). It had no important US application until 1914, when it was adopted by engineers working on the Panama Canal to indicate the positions of locks, gates, fender chains, and water levels on a miniature lock system. BuOrd bought a test motor from General Electric in 1918, as soon as it learned of this system. Mindell, p 48. The major developmental problem still to be solved was to achieve sufficient precision. In March 1921 Lieutenant William R Furlong, head of the Fire-Control Section (later chief of BuOrd) visited Germany to see the wartime fire-control system, which used synchros. The Germans had adopted small-diameter motors. Most of those made were rejected as insufficiently precise. Furlong argued that the larger-diameter GE motors were easier to make to the desired degree of precision.

42 In August 1921 GE received contracts for the other two *Maryland* class battleships (one later cancelled), for six *Omaha*-class light cruisers, and for twelve destroyers. The *Maryland* system was completed and tested in June 1923. Correspondence about the press release is in folder 17-686-2619 of 1923, RG 74 series, NARA. Arma produced an alternative selsyn system, and Ford produced a DC selsyn for anti-aircraft batteries (it could be installed on board older ships with DC systems). The Arma system went into the two *Lexington*s. The Ford system was for the anti-aircraft batteries of the carriers and three battleships. No previous DC system had been self-synchronous. The trick was to induce currents in the windings of a stator by spinning a magnetised rotor. Three sets of windings 120 degrees apart had different currents induced in them. The receiver had a similar spinning rotor, and the currents in its stator windings depended on its own position. When the two matched, no current flowed from transmitter to receiver; otherwise current flows moved the two stators into the desired position. The Arma system differed from GE's in that it used a more elaborate rotor with 360 bars, hence it was precise to within a degree in its positioning. A second set of brushes was geared 36:1, so that each step was equivalent to 1/36 degree, which was precise enough for fire control. By this time there were also electric equivalents of differentials, which could be used, for example, to keep an indicator set to zero if it was properly aligned. Lieutenant Commander C H Jones, BuOrd, 'Modern Fire-Control Installations,' lecture for the Naval Postgraduate School, 23 February 1926, RG 38 ONI files, NARA. About 1930 BuOrd decided to standardise synchros to provide mechanical and electrical interchangeability among all instruments. The clearest indication of such standardisation was that in 1944 the cruiser *Pensacola* had both a GE (Mk 22) and an Arma (Mk 18) director. Note that US synchro systems generally used two separate synchros, coarse and fine (vernier), geared together (1:72 ratio). The high-speed synchro acted as vernier to the low-speed one. One revolution of the low-speed synchro was 360 degrees; one of the vernier was five degrees.

43 Work on a stable vertical began at the Bureau of Standards during World War I, and it was approved for production about 1921–22. According to Wright, 'Questions… Pt 1,' the first was tested on board USS *Nevada* between 5 and 10 May 1918. The 1923 edition of the BuOrd *Confidential Bulletin* (p.5) reported initial installations aboard USS *Nevada* and USS *Tennessee*, with more to follow in quick succession. The manual for the Mk IX director (stable vertical) is dated June 1924 (OP 397). According to Mindell, p 55, Arma developed the first fully effective stable vertical (Mk 26) in 1929 (ie, the first useful for cross-level). It was put into production as Mk 29 for the *Portland* class. Through the *Helena* class, cruisers had the Arma Mk 30. Modernised battleships had Mk 32. The wartime standards were Mks 41 (Arma) and 43 (GE).

44 Trent Hone, 'Evolution of Fleet Tactical Doctrine in the US Navy, 1922–41,' *Journal of Military History 67* (October 2003), pp 1107–48. By about 1930 there were battle plans for normal and reverse action at each range band, plus special plans for other situations. BuOrd wanted to increase the elevation of the guns of the first three classes the US Navy modernised (*Florida, Arkansas* and *New York* classes), but the project was dropped due to strenuous British objections. (Ironically, the idea had been raised partly because of rumours that the British were increasing their own battleship gun elevation). The range of these ships' guns was therefore limited to 21,000 to 23,000 yards. By 1927 the British no longer opposed increased elevation, and it was provided in the *Nevada* and later classes. Presumably the official view that Hone reports was modified as modernised battleships with long-range capability entered the fleet. Even some unmodernised ships had extreme-range capability: during Fleet Problem X (1930), *New Mexico*, with the same main battery as *California*, opened at 35,000 yards. Citing US Naval War College material, Hone shows that the US Navy of the 1930s was unaware of both the extreme ranges the Japanese were achieving by reconstruction (eg, 37,000 yards for 14in/50s) and their increased speeds. Much the same could be said of the interwar Royal Navy.

45 Figures from Hone, 'Evolution of Fleet Tactical Doctrine' (see note 44).

46 BuOrd *Confidential Bulletin* for 1933, p 32. Ships for which installation was *not* authorised were the oldest surviving battleship, *Arkansas* (whose sister *Wyoming* had been demilitarised as a training ship); the three *Idaho*s (being modernised); and the two *Californias* (of which *Tennessee* had the new prototype). The new stable vertical director was Mk 32, the first to be fixed on a ship's deck. Previous stable verticals trained in the direction of the target, Mk 30 (*New Mexico* class) being the last.

47 Mk 4 was incorporated in the first US anti-aircraft director, Mk 19. In the modernised *Idaho*-class battleships and the first *San Francisco*-class cruisers it was replaced by Mk 28, a synchro system with a Mk 9 range-keeper. Mk 33, a parallel design for destroyers with a Mk 10 range-keeper, superseded it. Mk 33 was designed to control dual-purpose 5in/38 guns rather than the single-purpose 5in/25s of the earlier systems.

48 The proposal for the new type of fire-control system was signed by H H Willard for GE and by H C Ford for Ford Instrument. The concept was diluted, particularly after Willard, its architect, died, according to a 30 November 1935 letter from the Special Board on Naval Ordnance to the Chief of BuOrd referring to 10 November recommendations by T section, in RG 74 correspondence in NARA II. Critics claimed that BuOrd should follow past practices, but the letter showed that there were no truly consistent lines of development to follow, hence that it was sensible to seek new ideas.

49 In the secondary batteries of the three *Maryland*s (the only synchro battleships), and in the main batteries of the new 8in ships (*Lexington*-class carriers and *Pensacola*-class cruisers), in six light cruisers, and in twelve destroyers.

50 The horizontal gyro could point towards the target or it could lie in an arbitrary direction, against which angles could be measured. General Electric chose the former, and was assigned primary responsibility; Ford developed the latter as back-up. There was some question of how the gun-train order would be generated. Existing systems, such as the ones in the battleships then being modernised, used the director as a target designator. The computer added deflection and cross-level corrections, and the result was transmitted to the guns. This technique was covered by a Sperry patent. To avoid the patent, GE and Ford both proposed a director prism set automatically with corrections, the result going directly to the guns. It was rejected as too complicated. At the BuOrd conference examining the new system, GE asked that it receive a proprietary contract in which Ford would be subcontractor. BuOrd (RG74) S71 correspondence file for 1926–42, NARA.

51 Based on an account in *Fire Control Installations 1934*, notes prepared for the Postgraduate School, US Naval Academy, pp 99–108, in Navy Yard Library.

52 Mod 0 controlled 5in/51 guns on board the rebuilt *Pennsylvania*s, and Mod 3 controlled them on board the modernised *New Mexico* class. Mod 2 was an auxiliary 8in director for the *San Francisco* and *Portland* classes. Designations from the 1933 edition of *Gunnery Instructions*. The 1945 catalogue of BuOrd equipment lists Mod numbers up to 21, some of them controlling 6in/53s (*Omaha* class).

53 *Brooklyn*s were built with Mod 0 and Mod 1 directors (Mod 1 aft). The heavy cruiser *Vincennes* (CA 44) had the first 8in versions, Mods 2 and 3; the OD for this version (1936) clearly refers to a spotting glass. Some or all *Brooklyn*s were completed with 9ft Mk VIII spotting glasses, which were transferred in 1939 to *Omaha*-class cruisers (C&R S71 correspondence files beginning in December 1938 include discussion of where the spotting glass should be mounted). The rangefinder figures in the draft Mk 34 Mod 1 manual, dated 11 September 1939 (applicable to CL 40–43 and to CL 46–48) in the RG74 (S71) NARA correspondence file for 1926–42 (an OD version is in the OD file at College Park). *Main Battery Gunnery Notes, Light Cruiser, Brooklyn Class*, issued by Cruisers, Battle Force, 9 January 1941 (NARA II) describes only the version with an 18ft stereo rangefinder. The two high turrets had stereo rangefinders, the two low ones coincidence. As redesigned Mk 34 carried eight personnel: at the rear were the original ones: pointer, trainer, spotter and radio man. The rangefinder required a range-reader. At the front were the cross-leveller and two sight-setters. The sight-setters set sight angle and sight deflection from the range-keeper or auxiliary computer in plot (in primary control, only sight deflection). The original versions with the Mk 8 spotting glass weighed 4500lb. The version for an early *Cleveland*-class cruiser weighed 7500lb. By 1945 there were 29 Mods, including two for 14in/50 (Mods 24 and 25) and two for 16in/45 (Mods 26 and 27). The others were for 6in/47 and 8in/55 ships. Some of this material is from the BuOrd World War II fire-control history.

54 Based on a description in a 27 February 1936 letter from George A Chadwick, then at Puget Sound Navy Yard, to Rear Admiral H H Stark, then Bureau chief. Chadwick cited assistance from numerous battleship gunnery officers in formulating his description. Destroyer officers were resisting moving the range-keeper below decks because Mk 33 was proving successful, and because they did not realise how difficult it would be to use an open director in cold weather (the successor Mk 37 was entirely enclosed, its computer below decks). S71 files, BuOrd correspondence 1926–42, NARA. Chadwick cited initial memos by Commander Jones dated 15 November and 31 December 1935.

55 During World War II, plans called for installing enclosed directors in the *Pensacola*-class cruisers. They and the *Northampton*s were modified in 1940, their open 8in directors and their after anti-aircraft directors being moved onto stub mainmasts just forward of No. 3 turret. In this position the 8in director suffered from blast from the turret, particularly in the *Pensacola*s, where it was in the superimposed position. The solution ordered in 1942 was to fit all these ships with Mk 34 directors. That was done in the *Northampton*s, but in the spring of 1945 it was suddenly realised that the *Pensacola*s could not take so much topweight. Instead they were assigned Mk 35s from destroyers refitted with dual-purpose guns. The *San Francisco* class also could not accommodate the heavy Mk 34, and plans also called for them to receive Mk 35s (none was ever so refitted).

56 According to *Progress in Gunnery 1924* (ADM 186/263, issued April 1925), USS *New York*, which had a spread of 1000 yards in the Grand Fleet, had cut that to 350 yards. The *Maryland* class (16in ships) had spreads of only 350 to 400 yards, but the *Idaho*s were in serious trouble, with spreads of 900 to 1000 yards.

57 US battle line tactics are described by Trent Hone, 'Evolution of Fleet Tactical Doctrine in the US Navy, 1922–1941,' *Journal of Military History 67* (October 2003), pp 1107–48. Hone points out that possession of battlecruisers by both Britain and Japan was a cause of particular concern to the US Navy. Much effort went into devising tactics that could counter them. Reversing course would place the enemy van opposite the US rear, in a poor position to attack. It would also move the enemy's battlecruiser force into an unfavourable position. The official papers Hone cites do not mention the range-rate issue. Hone describes the moderate-range battle plan developed in 1941 by Pacific Fleet commander Husband E Kimmel, which included reverse action to concentrate firepower on the rear of the Japanese battle line and, in effect, isolate the battlecruisers assumed to be at its head. A 1934 'countermarch' (ie, reversal) exercise was much publicised.

58 *Gunnery Instructions*, 1933 edition, as reprinted with amendments in 1943. I am grateful to Trent Hone for comments on the US rocking and British zigzag concepts.

10 The US Navy at War

1 This account relies heavily on Trent Hone, '"Give Them Hell!": The US Navy's Night-Combat Doctrine and the Campaign for Guadalcanal', *War In History* 13 (2), pp 171–99.

2 The first series of post-World War I destroyers (*Farragut* class) were limited to two sets of torpedo tubes. In the next (*Mahan*) class, adoption of high steam conditions made it possible to fit a third set. The fourth was added at the expense of one 5in gun. The *Sims* class (1939) cut back initially to three sets of tubes, to restore the fifth 5in gun (the role of the destroyer was reassessed to emphasise fleet air defence) and then back to two due to gross overweight. The next class (*Benson*) adopted quintuple tubes, which were standard for US war-production destroyers.

3 As described by Hone (see note 1 for full source details), based on contemporary tactical publications (1937–41).

4 Cruisers always engaged in night main-battery practice, because in combat they would have worked

with destroyers either making or repelling night attacks. When the heavy cruisers appeared, they naturally engaged in night practice, and this continued after they were split into a separate category. US Navy Reports on Gunnery Exercises.

5 Despite its designation, which implied an in-service date of 1933, the Mod 2 service version of the 'Long Lance' became operational in 1936. This date, which coincided roughly with that of the last Japanese Grand Manoeuvre observed (covertly) by the US Navy, may explain why the elaborate Japanese night-attack techniques were unknown to it. Mod 2 had a range of 20,000 metres (21,872 yards) at fifty knots and carried a 490kg (1080lb) warhead.

6 *Helena* detected the three Japanese groups at 27,000 to 30,000 yards and deduced enemy course and speed. Of the three surviving cruisers, only she reported using radar fire control, engaging a target at 4200 yards with a 200-yard rocking ladder under automatic control (one director for train, another for elevation) and rapid continuous fire (in two minutes she fired 175 rounds). She had tracers, and her spotter reported that she was perfect for deflection and that she was making numerous hits. She was hitting a ship firing at *San Francisco*. Later she fired at a target at 8800 to 9400 yards in full automatic (125 rounds expended) and at another at 16,400 yards (about sixty rounds fired in one minute of rapid continuous fire). It is not clear which ships, if any, were being hit. The Japanese main body consisted of the fast battleships (battlecruisers) *Hiei* and *Kirishima*. The screen consisted of eleven destroyers led by one light cruiser. As an indication of the quality of intelligence (and of the confusion of a night action), the Pacific Fleet summary (copy in ADM 199/1358) describes a Japanese force consisting of a northern group of a light cruiser and four destroyers (probably correct), a middle group of two battleships (not in close formation), followed by a heavy cruiser (incorrect) and three destroyers; and a southern group comprising one heavy and one light cruiser and three destroyers (incorrect). The after-action report includes numerous references to torpedo hits on the battleships and cruisers (none seems to have been made) and to Japanese destroyers blowing up (which did not happen). One Japanese destroyer was sunk that night and another, disabled, was sunk by cruiser fire the next morning.

7 Morrison, Rear Admiral Samuel E, *The Struggle for Guadalcanal, August 1942–February 1943* (Vol V of the *History of United States Naval Operations in World War II*; Boston: Little, Brown, 1949 [reprint 1975]), p 249, describes how USS *Monssen*, turning on recognition lights after being illuminated by starshell, was instantly lit by Japanese searchlights and then destroyed. The idea of recognition lights seems to have come from the Germans, who began using them before World War I.

8 Charles Haberlein of the US Naval Historical Center realised what was happening when he studied the ship's TBS (tactical radio) log in connection with his own work on the Guadalcanal Battle. Mr Haberlein was the historian accompanying the Ballard expedition to Guadalcanal, responsible for identifying the warship wrecks they found. His expertise in recognising various ship details (he is Curator of Photographs at the Naval Historical Center) made him invaluable to that expedition.

9 Hits counted before *Atlanta* sank the next day all showed green dye, the color used by *San Francisco*. However, *Atlanta* claimed that she had been hit by three or four salvoes, whereas *San Francisco* reported firing only two before ceasing fire. *San Francisco* may have taken *Atlanta* under fire, shifted away, and then returned to her, where her after-action report showed two distinct targets. The after-action report shows seven main battery salvoes fired at the initial destroyer or small cruiser target, which was illuminated by starshell at 3700 yards – and, presumably, misidentified. *San Francisco* fired 160 rounds of 8in altogether. Morrison, *Guadalcanal*, p 247 (see note 7 for full source details), notes that in November 1945, Lieutenant Commander Bruce McCandless, who took command of *San Francisco* after the hit on the bridge, told him that he was engaging a Japanese ship beyond *Atlanta*, and that low-trajectory rounds may have gone through *Atlanta* en route to that target. By this time *Atlanta* was dead in the water due to one or two torpedo hits.

10 As suggested by Admiral Nimitz, Pacific Commander, in his report on the action.

11 *San Francisco* range was 2200 yards; *Portland* range was 4200 yards. *San Francisco* claimed at least eighteen hits and *Portland* claimed four hitting salvoes (twenty-four rounds, out of a total of ninety rounds of 8in AP shells she fired that night) from her forward 8in guns at 4000 yards. *Portland* found the range using her Mk 3 radar. She used only her forward guns because she had just suffered a torpedo hit aft, which sheered off propellers, jammed her rudder, and jammed her after turret in both train and elevation. The following morning, disabled and steaming in circles, *Portland* still managed to sink the damaged Japanese destroyer *Yudachi* with six six-gun salvoes at 12,500 yards – a performance described as a highlight of the action.

12 Shrapnel from Japanese bombardment shells killed many exposed topside personnel who would have been safe from AP rounds. *San Francisco* reported fifteen major-calibre hits and many lesser ones; at one time twenty-five separate fires were burning on board. *Portland* reported two 14in hits, neither doing much damage. A torpedo hit aft sheered off propellers and left her with little control.

13 Hone, 'Give 'Em Hell….' (see note 1 for full source details).

14 The BuOrd (RG 74) correspondence series for 1926–72 includes, in its S71 section, a 20 January 1938 Bureau of Engineering memo on the use of 'centimetre waves' for fire control, replying to a 31 December 1937 BuOrd letter, and commenting on an earlier one (30 March 1937): 'everything possible is being done to expedite development of radio detection and ranging. Although shipboard equipment is not yet available (power output is still too small), it is expected within the next two years.' The same file includes a 30 March 1935 letter from the Bureau of Engineering confirming a 1933 letter from the Naval Research Laboratory explaining the potential of what were then called Micro Rays, and describing laboratory progress (eg, messages transmitted 200ft and reflection at similar ranges).

15 Details from *Fire-Control Equipment: Fire-Control Radar Types FC and FD–Operation* (ORD-657), February 1942, NARA II. Compared to FA (Mk 1), FC had a more powerful transmitter (40 compared to 15 kW) and a much better indicator, with a dial marked in ten-yard increments (an FA operator had to estimate range in hundreds of yards from his scope). The manual noted that British experience was that radar range was more reliable than optical range, eg, when HMS *Hood* engaged *Bismarck* the optical range was 23,000 yards and the radar range 27,000; the opening salvo fell about 4000 yards short.

16 Commander Cruisers Pacific Fleet, *Gunnery Doctrine and Standard Fire-Control Procedures, Supplementing PAC-10*, August 1943.

17 *Washington* reported firing forty-two rounds of 16in, sinking the target. She reported that the US force sank one large cruiser or battleship (sunk by *Washington* on her own), two large cruisers (sunk by *South Dakota* and *Washington*), and one destroyer. USS *Washington* had damaged one 14in battleship (silenced and out of control), left one destroyer burning, and silenced between five and nine light craft. The sinking claims were erroneous except for the destroyer, and no light Japanese craft were present; the 14in battleship reported silenced was *Kirishima*.

18 Starshell was fired short, possibly because the secondary director involved did not take account of the target range rate. It blinded some of the spotters.

19 Brad Fischer, who co-wrote a two-part article in *Warship International* on the gunnery accuracy of US fast battleships in World War II, has found evidence that *Washington* made many more hits. The original claim of nine hits apparently originated with the anti-aircraft fire-control officer of *Kirishima*, Lieutenant Commander Horishi Tokuno, whose statement can be found in the US Strategic Bombing Survey *Interrogations of Japanese Officials*. Fischer has found what he considers more reliable Japanese sources suggesting that the ship took twenty 16in hits and seventeen 5in hits. Private communication by Brad Fischer; this material is not in his articles.

20 Details from *Fire-Control Equipment: Fire-Control Radar Type Mk 8 – Operation*, January 1943, NARA II. Calibrated range was 45,000 yards, with an accuracy of fifteen yards plus or minus 0.1 per cent of measured range, eg twenty-nine yards at 14,000 yards; the beam swept a thirty-degree-wide area, with an accuracy of six minutes of arc (0.1 degree). Targets could be detected out to 60,000 yards, but maximum instrumented range was 45,000 yards. In fire-control mode, the thirty-degree sector was scanned ten times per second (there was also a target-acquisition mode, in which a two-degree sector was scanned. A battleship could be detected at 35,000 to 45,000 yards, and a 16in splash at 20,000 yards on the B-scope. Mk 8 was in effect phase-scanned, its array static while a phase-changer moved behind it. Mk 13 was a more conventional, rapidly scanned dish inside a radome.

21 This account is based largely on US after-action reports provided to the Royal Navy, and collected in ADM 199/1498; it is supplemented by the account in Morrison, Rear Admiral Samuel E, *Leyte, June 1944 – January 1945* (Vol XII of the *History of United States Naval Operations in World War II*; Boston: Little, Brown, 1958 [reprint 1974]).

22 According to the Battle Division 4 report, she fired on a target distinct from the one being engaged by *West Virginia*. Morrison, *Leyte…*, says that she used *West Virginia*'s splashes as her aim point.

23 BuOrd World War II Fire-Control volume (V), 29–32. It refers to vulnerability.

24 This account is largely based on US after-action reports collected in ADM 199/1330.

25 Lacroix, Eric, and Wells, Linton, *Japanese Cruisers of the Pacific War* (Annapolis: Naval Institute, 1997), p 316. Capacity was 1200 shells (overload was 1260).

26 Lacroix and Wells, p 316, report no 8in hits on *Nachi*, based on Japanese documents. All five hits were described as 5in, but were probably from 8in from *Salt Lake City*: three to starboard at 0350 Japanese time (presumably 0850 in US terms, which does not correspond to claims by the US cruiser), one at 0648 (presumably 1148), and one at an unknown time. The fourth hit jammed No 1 turret. One of the first three hits struck the bridge and damaged fire-control circuits, killing eleven and wounding twenty-one on the bridge. Repairs took about a month. *Maya* was not hit at all.

11 The Japanese Navy

1 Lacroix, Eric, and Wells, Linton, *Japanese Cruisers of the Pacific War* (Annapolis: Naval Institute, 1997) p 769. As drawn on p 103, the Type 14 director resembles the Royal Navy type, with separate pointer and layer telescopes.

2 Maximum range of the initial version was 19,000 metres (20,778 yards). A modified Type 13 was formally adopted in 1924, although it had been in service for some time. Type 14 (ie, 1925, adopted 1927) could be used for both low- and high-angle fire. The replacement, standard in wartime, was Type 94 (ie, 1934), tested on board the battleship *Nagato* (formally adopted 9 October 1934). It carried a binocular spotting glass through which the control officer observed the target. There were separate pointer and trainer telescopes, and a separate cross-level telescope. The other two personnel were a talker and a sight-setter. Versions of the Type 94 director: Mod 1 for battleship main batteries (152mm (6in) binoculars), Mod 2 for battleship secondary batteries and cruiser main batteries (120mm (4.7in) binoculars), Mod 3 for destroyers (120mm (4.7in) binoculars), and Mod 5 for *Agano*- and *Oyodo*-class cruisers (120mm/4.7in binoculars). A total of 132 directors were made for the whole Japanese fleet. The *Yamato* class had Type 98 directors. *Ban* was a board. *Hoi* was bearing angle; *Shageki* was firing. Directors were called *Hoiban* and calculators were *Shagekiban*.

3 The Japanese designated equipment by the year of adoption. They used either a date within the reign of the current Emperor or a date (last two digits only) within the Japanese calendar, in which the Western year 1940 was 2600 (so that the fighter adopted that year became Type 00, typically Type 0 – the famous Zero). Designations by year are sometimes misleading, because a device could enter service without any such designation, one being applied only later – indicating a later year. Equipment adopted during World War I was designated in a series beginning in 1912, so a 1914 device was Type 3 (this reign ended in 1926). For the next reign, which was Hirohito's, the calendar system had largely been adopted, so Type numbers ran to high double digits. e.g. Type 89 for 1929 or Type 93 for 1933 (however, in other contexts the Japanese continued to use the year of the reign, so that 1945 was the nineteenth year.

4 These devices are named by Lacroix and Wells, *Japanese Cruisers*, p 770. The equivalences to British equipment are the author's. The Japanese apparently did not receive any form of the Dreyer Table. Drawings of fire-control equipment in Japanese books show range plotting boards, but not Dreyer's trademark measuring grid.

5 *Progress in Gunnery* 1925.

6 Barr & Stroud ancestry (and adherence to that pattern) are from a statement by the Chief Engineer of Aichi Clock quoted in Report O-31, 'Japanese Surface and General Fire Control,' of the US Naval Technical Mission to Japan, dated January 1946. Terminology in the report suggests British authorship, eg, the computers are all called tables. The first Japanese prototypes were Type 91s for 140mm (5.5in) guns, the first being tested on board the cruiser *Kiso*, then a training ship. A modified version was adapted for heavy guns (204mm/8in, 356mm/14in and 406mm/16in) as Type 92. Twenty-one were made beginning in 1933

(final delivery was in 1943). Mod 1, for the *Mogami* and *Tone* classes, had a built-in mechanism to compute target course and speed, requiring only an inclinometer input. Delivered in 1937 for the 155mm (6.1in) guns of these ships, Mod 1 was adapted to 204mm (8in) guns when the ships were rearmed in 1939. This computer incorporated a wind-correction mechanism. The *Yamato* class had a further improved Type 98 computer. Installation data from Lacroix and Wells, *Japanese Cruisers*, p 770. The layout of the Type 92 table (computer) is shown in Lacroix and Wells, p 234, with personnel positions.

7 Lacroix and Wells, *Japanese Cruisers*, pp 235–6. Their table of mean salvo spreads shows 280 to 330 metres (306 to 361 yards) for the *Myoko* class at 20,000 to 22,000 metres (21,872 to 24,059 yards) in 1936, and 380 metres (415 yards) for all ten-gun ships in 1940; it is not clear that the two special salvo-limiting devices had much effect.

8 Trent Hone, '"Give Them Hell!": The US Navy's Night-Combat Doctrine and the Campaign for Guadalcanal', *War In History* 13 (2), quoting comments on Cape Esperance by Admiral Nimitz, based on the battles of Savo and Cape Esperance. Nimitz thought that US cruisers, firing on the basis of radar bearings and ranges, began hitting on the first salvo. We now know that this was not always true, as targets considered sunk were sometimes merely leaving the radar-range gate.

9 Report O-29, 'Japanese Fire Control,' of the US Naval Technical Mission to Japan.

10 According to Lacroix and Wells, *Japanese Cruisers*, p 772, until World War I Japan imported Barr & Stroud rangefinders, but afterwards they were produced by the Nippon Optical Manufacturing Company of Nagoya. The first, Type 5, was a 4.5m (14.7ft) unit designed in 1916 and completed in 1917; it was fitted on board the battleship *Yamashiro*. A 10m- (33ft-) base instrument (1918) was fitted in 1921 as Type 7 on board the battleship *Nagato*. An 8m (26.2ft) duplex instrument was adopted in November 1923 as Type 13; it was fitted to the battleship *Haruna* in 1923. Further rangefinders with bases of 2.5m, 3m, 4.5m, and 8m (8.2ft, 9.8ft, 14.7ft and 26.2ft) were Types 89 (1929), 90 (1930), and 93 (1933). *Yamato* and *Musashi* had 15m (49.2ft) base rangefinders.

11 Notes on shells from Lacroix and Wells, *Japanese Cruisers*, pp 758–60 and Report O-19, US Naval Technical Mission to Japan, on Japanese Projectiles. All Type 91 projectiles, from 152mm (6in) up to 457mm (18in), were flat-headed for stable water travel. Each squadron (*Sentai*) had its own dye colours, eg in *Sentai* 3 *Kongo* was red, *Haruna* black, *Kirishima* blue, and *Hiei* uncoloured.

12 This secrecy largely negated it, because the relevant tactics could not be practiced in peacetime. The shell would only be effective at long range, about 20,000metres (21,872 yards) for battleship guns and 18,000metres (19,685 yards) for 204mm (8in) cruiser guns. David C Evans and Mark R Peattie, *Kaigun: Strategy, Tactics, and Technology in the Imperial Japanese Navy, 1887–1941* (Annapolis: Naval Institute Press, 1997), pp 264–5.

13 Eg, the hit by HMS *Prince of Wales* which entered a fuel oil tank on board the German *Bismarck* and, by costing her a great deal of fuel, forced her to turn towards France and, ultimately, destruction.

14 Evans and Peattie, *Kaigun*..., pp 131–2. By 1907 Japanese strategists argued that their naval strength should be predicated on that of the United States, because whatever current relations, the United States had the greatest potential to harm their country. They calculated that, to defeat a US attack, Japan needed a fleet 70 per cent as powerful as that of the United States. This figure was predicated on the widespread assumption that a fleet needed a 50 per cent margin to win, and that the US fleet would be weakened by the effort of crossing the Pacific (it was widely assumed that a fleet lost fighting power as it moved from its base, as the Russians had certainly found at Tsushima). Once stated, the 70 per cent figure became a guiding dogma for the next three decades. Evans and Peattie, p 143. The 70 per cent figure was linked with plans, made between 1907 and 1922, for an 'eight-eight' fleet (eight modern battleships and eight modern armoured cruisers, later replaced by battlecruisers), where modern meant less than eight years old.

15 Evans and Peattie, *Kaigun*..., p 250.

16 Description of concentration tactics from *Progress in Gunnery 1936*, (ADM 186/338), p 95. The numerous references to Japanese radio and spotting procedures strongly suggest that the report describes exercises, and that the British (like the US Navy) were reading Japanese codes.

17 At this time the Royal Navy was experimenting with extreme ranges, the *Nelsons* having fired at up to 37,000 yards, and *Hood* and heavy cruisers at up to 27,000 yards. Given the reduced probability of hitting at long range, the British credited reports that ships were carrying more ammunition, up to 150 rounds per gun in battleships and perhaps 200 in cruisers, compared to the usual 100 in other navies. In fact a typical ammunition load for a ten-gun heavy cruiser was about 1300 rounds.

18 Figures from Evans and Peattie, *Kaigun*..., pp 260–2.

19 Work on oxygen torpedoes, begun in 1917, was stopped after explosions, but resumed in 1928 due to reports that the British had 24.5in oxygen torpedoes on the battleships *Nelson* and *Rodney*. Ironically, the British had themselves abandoned oxygen torpedoes due to explosions. The Japanese persisted, and in 1933 they tested a successful oxygen torpedo. It was adopted on 28 November 1935 as Type 93 Model 1 Modification 2; 1150 of these 'Long Lances' were made. They entered service on board heavy cruisers in 1938 and on board modern destroyers, beginning with the *Kagero* class, in 1940. Although it was known that the Japanese had 609mm (24in) torpedoes, there was apparently no inkling of their capabilities until late 1943. Performance: 20,000 metres (21,872 yards) at forty-eight knots, 32,000 metres (35,000 yards) at forty knots, 40,000 metres (43,744 yards) at thirty-six knots. Model 3 of 1943 was derated to 15,000 metres (16,400 yards) at forty-eight knots, but range at the other speeds was unchanged. Type 93 superseded a compressed-air Type 90 (1933) of the same dimensions but slightly lighter. It made only 7000 metres (7655 yards) at forty-six knots, 10,000 metres (10,936 yards) at forty-two knots, and 15,000 metres (16,400 yards) at thirty-five knots, figures not too different from its Western contemporaries. The US Navy seems not to have been certain of the characteristics of the 24in torpedo until some time in 1944. Torpedoes were recovered at Guadalcanal in 1942, and their markings indicated a maximum-range setting of 20,000m (21,872 yards); however, BuOrd doubted that they were oxygen weapons in a November 1943 report. The following January 3rd Fleet distributed a 'reliable' report on the *Shimakaze* class including torpedo ranges (30,000 metres (32,808 yards) at thirty-two knots, 20,000 metres (21,872 yards) at forty-two knots, and 10,000 metres (10,936 yards) at forty-six knots), but a Japanese notebook recovered at Saipan that June showed a range of only 10,000m (10,936 yards) at thirty-eight knots. In July 1944 ONI distributed a handbook of the Japanese fleet (its first since December 1942) announcing the long-range torpedo as news. Data from a BuShips file (RG 19) on foreign warships in NARA 2.

20 Evans and Peattie, *Kaigun*..., pp 266–73. The midgets were unusual for their time for their high speed: nineteen knots for fifty minutes underwater (28,990 metres/31,700 yards); surface speed was six knots (thirteen-hour endurance). Each carried two 450mm (17.7in) torpedoes. Thirty-six were built between 1936 and 1940, the prototype having been built in 1934; four tenders were produced, two laid down especially for this purpose. Each could carry twelve midgets.

21 Evans and Peattie, *Kaigun*..., p 273, particularly cite the Type 88 Mod 1 adopted in 1932.

22 Evans and Peattie, *Kaigun*..., pp 277–81.

23 Evans and Peattie, *Kaigun*..., pp 282–7, based on fragmentary sources, hence probably not corresponding precisely to any one battle plan. They cite Japanese estimates that the initial mass torpedo attack should cripple or sink at least ten US capital ships. Each destroyer division (four ships) should account for one capital ship.

24 Evans and Peattie, *Kaigun*..., p 293, ascribes the lack of testing to both a perennial shortage of fuel and the absence of many of the vital material components of the plan, such as sufficient numbers of Type 93 torpedoes. Exercises tested phases of the plan, but never the full cycle of twilight, night, and day actions.

25 It probably helped that torpedo developers in the United States and United Kingdom refused to imagine that the Japanese could do better than them. This point is made by John Prados in his analysis of wartime naval intelligence, *Combined Fleet Decoded: The Secret History of American Intelligence and the Japanese Navy in World War II* (New York: Random House, 1995), particularly pp 31–2. A 1943 British translation of a Japanese report of torpedo lessons learned showed no inkling of the nature of the Type 93 torpedo (ADM 1/12647).

12 The French Navy

1 Much of this chapter is based on M P Peira, *Historique de la Conduite du Tir dans la Marine* (Paris: Memorial de l'Artillerie Francaise, 1955). Peira was a senior French fire-control engineer with direct experience of many of the systems involved. His book also gives considerable insight into the more general fire-control problem.

2 SHM (Service Historique de la Marine), BD 16, 'Conferences sur la Tir a Bord,' 1899, lectures delivered by a Lieutenant de Vaisseau Freund at the Ecole des Cannoniers (gunnery school). They seem to concentrate on a sophisticated analysis of firing errors, although the problem of a moving target is also discussed (the writer differentiates techniques to be used for high and low range rates). No geometrical analysis of the moving-target problem seems to be provided. The French tried firing at ranges up to 4000 metres (4373 yards) in 1898–9, but they do not seem to have fired at 6000 metres (6560 yards) until 1903 or later. Information from John Spencer. In a 1931 War College essay, French Lieutenant de Vaisseau Deshieux (SHM, carton ICC 285) summarized French pre-World War I gunnery practice ranges. When systematic practice began in 1900–01, the set range was 3500 to 2000 metres (3827 to 2187 yards), despite Admiral Fournier's assumption of 5000-metre (5468-yard) range a few years earlier. However, during the 1903–04 exercises ships opened fire at 6000 metres (6560 yards), and in 1906–07 the battleship *Gaulois* opened fire at 7000 metres (7655 yards) (her average range was 6000 metres/6560 yards). In 1907–08 the battleship *Republique* opened fire at 7900 metres (8639 yards), and in 1909 she fired her secondary battery of 164mm (6.4in) guns at 10,500 metres (11,482 yards). The *Danton*-class semi-dreadnoughts opened their first gunnery practices (1911–12) at 10,000 metres (10,936 yards), and usually fired at 8500 to 9000 metres (9295 to 9842 yards). The first French dreadnought, *Courbet*, fired at 11,500 metres (12,576 yards) on 2 April 1914.

3 Campbell, *Naval Weapons of World War II*, p 280. The train and elevation motors of director and turrets ran synchronously. Moving a hand wheel at the director generated a current approximately proportional to the training velocity, and lasting until movement stopped. This current was fed into the armature of the motor at the turret. The operator corrected any errors by hand. Campbell quotes German sources to the effect that these systems were not altogether successful. In the cruiser *Foch*, remote control was for train only, the electric motor driven via Janney hydraulic gear, with speed (but not acceleration) controlled by hand from the director. The 7600-ton cruisers built late in the 1930s (*La Galissoniere* class) had both Granat transmitters and remote-power control, but never used the latter in shoots.

4 Peira Vol 2, p 15, considers it to the credit of the French naval industry and of the naval artillery arm that they persevered despite early failures and the destruction of designs (during World War II). Directors and rangefinders were stabilised.

5 By the mid-1880s the French were using the Leguol sextant, which measured an angle by superimposing two images of the mast of a target ship. It in turn was based on techniques described in Lord Douglas's *Treatise on Naval Gunnery*, the 1826 edition of which was translated into French. In the 1890s the French became interested in the US Fiske stadimeter. Information from John Spencer, based on French archival data. The most common device in service in 1900 was probably the Model 1892 Fleurais Micrometer (used with a circular disc to find the range). A February 1910 catalogue of Ponthus-Therode instruments is in ONI file R-3-a No 62, on French naval rangefinding to 1914. They were introduced into service in 1903, and as of 1910 the latest model was dated 1908. About 450 of the small type and 250 of the larger type were in service, with another 100 of the large type on order for the *Danton* class. About ten had been sold to Argentina. At a 20m (65.5ft) height the device could measure ranges between 600 and 8000 metres (656 to 8750 yards) (at 40m/131ft height, minimum range was 1000 metres/1094 yards). Twisting a nut to the mast height gave the range directly. According to the ONI report, the average error at 5000 metres (5468 yards) was fifty metres (fifty-five yards), and the stadimeter was considered comparable to a Barr & Stroud coincidence rangefinder. Compared to the US Fiske stadimeter, it offered more magnification (12X) and allowed the range to be read off without taking the eye from the target. At this time the Barr & Stroud rangefinder was the principal type in the French navy. The stadimeter would be used after the fixed instruments had been destroyed in the earlier stages of a battle. It would probably also be used to follow up the Barr & Stroud, beginning with the ranges taken by that instrument (hence not relying on exact knowledge of the enemy's mast or

other heights). The large instruments were supplied to officers controlling turrets and the batteries of medium-calibre guns, the smaller ones to those controlling anti-torpedo boat guns.

6 According to a 1925 fire-control treatise for the French naval war college (in SHM), out to 7000 metres (7655 yards) the triplex was accurate enough for hits on the first salvo. At 7,000 to 12,000 metres (7655 to 13,123 yards) it was insufficiently accurate, but it offered sufficient precision to be used to measure the range rate. Beyond 12,000 metres (13,123 yards) precision declined rapidly, and at 15,000 metres (16,400 yards) the rangefinder was no longer good enough for range rate. A table showed mean errors in accuracy (ie, in total range) and in precision: 100 and 60 metres (109 and 66 yards) at 10,000 metres (10,936 yards); 200 and 120 metres (219 and 131 yards) at 14,000 metres (15,310 yards); 400 and 250 metres (437 and 273 yards) at 20,000 metres (21,872 yards). At this time it seemed that future ranges would exceed 40,000 metres (43,744 yards). In July 1937 a fire-control lecturer pointed out that the 11m (36ft) rangefinders of the new cruisers had an error of 250 metres (273 yards) at 20,000 metres (21,872 yards), and that the 12m (39ft) units on the new *Dunkerque*-class light battleships would reduce that margin of error to 175 metres (191 yards). He considered both figures decisive advantages over the 300-metre (328-yard) error he associated with the 10m (33ft) rangefinder of the German 'pocket battleship' *Deutschland*. Note that the 1925 book rejected an earlier official claim that range averaging made the triplex equivalent to a much longer-base instrument about three times the length of any single part. In a 1937 lecture on current trends in naval artillery, a senior officer commented that at the desired range of 40,000 metres (43,744 yards) rangefinders were unfortunately less accurate than guns: 250 metres (273 yards) for a 12m (39ft) rangefinder as in *Dunkerque*, 200 metres (219 yards) for her guns; figures for destroyers, using 5m (16.5ft) rangefinders at 20,000 metres (21,872 yards) were 170 metres/186 yards and 100 metres/109 yards).

7 Report dated 18 July 1898 by Vice Admiral Humann. Information courtesy of John Spencer. Costs were 4075 francs for Germain, 11000 for Eng. Spencer points out that the French almost adopted the Care system of stepping motors between 1898 and 1900, but dropped it because it was 25 per cent more expensive than the Germain system. The Care system was fitted to three ships, probably the old *Courbet* and *Devastation* (which had mechanical order transmitters using bevel gears) and the coastal battleship *Bruix*.

8 It was difficult to modify the Germain system to indicate greater ranges accurately, because a given change in pressure could push the needle only so far. Ideas included adding a second set of manometers, modifying existing ones so that the needle could turn around twice, and adding a light bulb that would indicate 5000 metres (5470 yards) more range. In the end the choice was a larger-diameter manometer (250mm/9.8in compared to 180mm/7in diameter) with a longer scale (but read with less accuracy). The jump to 14000 metres (15,310 yards) was accomplished simply by changing the dial face. Data courtesy of John Spencer.

9 Much of this information is from John Spencer, based on French archival material. Peira considered L-A much the same as the Barr & Stroud and Sperry systems. In it, six stator electromagnets in the receiver were connected in pairs to the corresponding segments in the stator of the transmitter, a seventh being connected to the source of current driving the rotor of the transmitter. As the transmitter rotor moved, currents were excited in the stator, and they in turn magnetised the stator elements in the receiver, turning its soft-iron rotor (which was not connected to the current). Each movement of the handle produced one pulse, sufficient to turn the receiver rotor thirty degrees (twelve pulses per full turn), equivalent to a change of range of twenty-five metres (twenty-seven yards) or a change of bearing of one quarter of a degree. The motor lost track if data were transmitted at a rate below that corresponding to sixty rpm.(ie, twelve steps/second, or thirty seconds to move a pointer from 9,000 to 18,000 metres/9843 to 19,685 yards). An improved version appeared in 1912.

10 The Lafrogne clock was described in an ONI report dated 21 January 1907, replying to a 20 November 1906 query, in file R-3-a No 09-416; at that time it was still experimental, having been tested on board the cruiser *Amiral Aube*, and currently on board *Pothuau*. The earlier dates given here are from John Spencer, based on French archival material. Input into the clock was somewhat awkward, using five binary keys (16, 8, 4, 2, and 1, so that to enter eleven metres (twelve yards)/second the operator pressed 8, 2, and 1). Unlike the Vickers Clock (and its derivatives) with its variable-speed drive, this one changed speed using differentials set by the keys. Before the *plateau* had been adopted, the French, like the British, tried simply measuring the difference between two rangefinder readings, a method they called *telechronomique*.

11 NHB, NID supplement (dated 1911) to the July 1909 Foreign Ordnance report (NID 878), pp 204–5. These are corrections to Vol II (gunnery) of the 1909 report, which unfortunately could not be found (NHB has Vol I, which describes guns and gun mounts). This report includes details of fleet firing practices, but observes that details of the new fire-control calculators (eg, the *pendule* and the *plateau*) were kept secret, despite the semi-alliance relationship with France. The 1906 report, which is in a single volume, includes no fire-control details. Peira notes that the French navy lost control of its shipboard artillery to the colonial troops in 1901, regaining it only in 1909. It is not clear that this bureaucratic change had much impact on gunnery development, which progressed considerably in the interim. As a member of the naval-artillery service, Peira began his history of French naval fire control in 1909, and tended to slight earlier developments. On the other hand, according to ONI file R-3-a No 530 (12 January 1911), 'after many years' delay they have taken up the British system of training gun pointers and gun crews, and are now proceeding with the development of long range shooting, but have doubtless been influenced by the experience they gained many years ago before the advent of very accurate pointing, the great improvement in powder, etc.' The officers' gunnery course was six months ashore, then three months of practical gunnery on board *Tourville*, then three months of fire-control instruction on board *Pothuau*, then a month-long tour of ordnance establishments around France.

12 Based on details in the 1911 NID gunnery supplement cited in the previous note. In March–April 1909, for example, the Mediterranean Fleet fired at the anchored hulk *Tempete* at 8000- to 5000-metre (8750- to 5470-yard) ranges at a speed of eight to fourteen knots. Each ship calibrated her guns at 8000-metre (8750-yard) range before firing. In July 1909 the fleet practiced concentrating fire by divisions, using the secondary-ranging technique. The Northern (Atlantic) Fleet conducted a similar practice firing against the anchored hulk *Tonnerre* in April 1909 at ranges of 6800 to 5500 metres (7440 to 6014 yards) (speed fifteen to seventeen knots). It tried the alternative coordinated salvo technique of concentration firing. Further practice in October 1909 through September 1910 was at 8500- to 7000-metre (9295- to 7655-yard) range at a speed of thirteen knots.

13 A 16 May 1906 article in the *Moniteur de la Flotte* may mark the beginning of French efforts at concentration, inspired by Japanese tactics at Tsushima. It is in a US intelligence report (R-4-a No 08-379) describing July 1909 concentration experiments using groups of three ships, firing either in turn at five-second intervals (to make spotting practical) or with each ship finding the range by firing at her opposite number in the enemy group, then shifting to the concentration target using that range. The source was a published article dated 22 August 1909. According to Peira, concentration fire was not practicable pre-war because radio was not yet sufficiently reliable, and visual signals were limited to good weather.

14 According to John Spencer, however, contemporary French files are filled with references to spotting. They were well aware that rangefinders could be inaccurate. The practice firing was probably limited to medium-calibre guns, because the heavy guns on *Jaureguiberry* fired so slowly, probably one round every two minutes.

15 File (in translation) from Barr & Stroud archives, courtesy of Professor Jon Tetsuro Sumida.

16 According to Peira, wind across was first taken into account in 1914–15.

17 In his otherwise encyclopedic discussion of fire-control mechanisms, Peira devotes no space to integrators, but he shows what looks like a Pollen or Ford double-ball integrator (*variateur de vitesse*) on p 146 of his Vol I (item 6 in figure 78, showing various computing mechanisms). He shows a British-style link (as in the Dreyer Clock with bearing output) to multiply (or, differently arranged, to divide). Le Prieur's 1918 fire-control handbook (coded EGN. Oa7 in the French archive at Vincennes) shows no integrator. Instead, he used a flexible Bowden cable to connect his Dumaresq-equivalent to a tiltable arrow on a carriage. As the range-rate bar moved down, it pushed one end of the pivoted arrow so that the latter matched the range rate given by the Dumaresq-equivalent. The operator set the pencil on the carriage to the average range, and reset range to match spots ('bonds'). See his Fig 7 on p 19. Le Prieur also had an equivalent to the British Dreyer Corrector, to take account of own- and enemy-ship speeds.

18 SHM, Ministere de la Marine (Etat Major-General, 4th Section), *Appareils de Conduite de Tir Systeme Le Prieur*, 1918, supplemented by a description in Peira. The range averager used rubber bands to represent range readings. This element of the system seems not to have been given to the Italians, who averaged by eye.

19 Letter, G L Schuyler (naval scientific attaché) to Rear Admiral Ralph Earle, chief of BuOrd, 4 June 1919, courtesy of C C Wright.

20 Peira, Vol 1, p 56, claims that the Royal Navy admitted that transfer was a problem in its Vickers system.

21 According to Peira, Vol 3, p 21, the French first obtained Siemens systems on board ex-German destroyers (*Delage* class) they received as reparations, but other writers claim that all such equipment was stripped before ships were surrendered. The French found such equipment 'a revelation'. In 1925 they tested the Italian Girardelli system on board the old cruiser *Gueydon* against the two preceding French systems (Sauter-Harle and Breguet). It was dropped; when the French navy visited Milan, the Italians said that they too had found the Girardelli unsatisfactory and would abandon it.

22 The Granat system and its derivatives and the Chalvet system. The Granat system was adopted in 1924 and was still in service at the outbreak of war in 1939. Compared to the L-A system, it offered instantaneous reaction, and ease of installation (receivers and transmitters could easily be slipped in and out of a network).

23 The Vickers director was installed on board *Bretagne* on a bracket below her foretop during a refit at Toulon (12 June 1919–18 October 1920). As in contemporary British practice, the associated rangefinder was separate, in this case a 3.6m (12ft) unit atop the foretop. The other dreadnoughts all received French directors inside their foretop platforms. At least *Provence* had a 2m (6.5ft) rangefinder installed atop it (*Courbet* had a 4.5m (14.7ft) rangefinder after a 1923–4 refit). During a 1927–8 refit *Bretagne* was fitted with a new masthead director carrying a 4.5m (14.7ft) rangefinder; two more such rangefinders replaced the earlier triplex atop the conning tower. The new director had two parts: a lower one with a flat front carrying two prominent ports (for pointer and trainer) surmounted by a smaller cylinder carrying the rangefinder. A new director (carrying the same rangefinder) was installed during a 1932–4 refit. It was a much larger turret (carrying the rangefinder) atop a barbette. The control ports were relocated into the top of the director. The director was replaced yet again during a January–April 1939 refit. This time the single 4.5m (14.7ft) rangefinder was replaced by two separate 8m (26ft) stereo units (one above the other) in two cylindrical bodies, the upper one smaller than the lower. The other two ships had much the same modifications. Photos of *Lorraine* at Alexandria after 1940 suggest that the lower element had the control windows on a flat face. In contrast to the *Bretagne* class, in 1929 the 304mm (12in) dreadnoughts all received cruiser-type directors during refits conducted in 1927–9 (*Paris*), 1927–31 (*Courbet*), and 1929–31 (*Jean Bart*). Material on the dreadnoughts is from Robert Dumas and Jean Guiglini, *Les Cuirasses de 23,500 tonnes* (Outreau: Lela Press, 2005). In contrast to Peira, Dumas and Guiglini (who used French archival materials) claim that at least at the outset gun elevation was increased only from twelve to eighteen degrees. For the second-generation battleships I have relied heavily on Robert Dumas, *Les Cuirasses Dunkerque-Strasbourg, Richelieu, Jean Bart* (Bourg en Bresse: Marines, 2001). I also benefited from gunnery trials information on *Dunkerque* supplied by Alexandre Sheldon-Dupleix from the French naval archive at Vincennes. The captions describing the cruiser systems are based on a series of French books on cruisers: Gerard Garier and Patrick du Cheyron, *Les Croiseurs Lourds Francais Duquesne et Tourville* (Outreau: Lela Press, 2003); Jean Guiglini and Albert Moreau, *Les Croiseurs de 8000 Tonnes* (Bourg en Bresse: Marines, n.d.); Jean Lassaque, *Le Croiseur Emile Bertin 1933–59* (Bourg en Bresse: Marines, n.d., probably 1993); Jean Moulin, *Les Croiseurs de 7600 Tonnes* (Bourg en Bresse: Marines, 1993); and Jean Moulin and Patric Maurand, *Le Croiseur Algerie* (Bourg en Bresse: Marines, n.d., probably 2002). The identification of ports with those in a British DCT is my own, based on the similarity in numbers and locations of ports. Material on super-destroyer and

Glossary

AC Aim Correction, Pollen's term for his fire-control system.

AFCC Admiralty Fire-Control Clock.

AFCT Admiralty Fire-Control Table.

AMC Armed Merchant Cruiser.

Analytic position-keeping Position-keeping based on deduction from observations (ie, analysis).

APC Armour-piercing, capped (shell).

Argo Clock Pollen's fire-control computer.

ARL Admiralty Research Laboratory.

BatDiv Battleship (or Battle) Division, US Navy term for a battleship unit.

Battle tracer Sperry device for automatic plotting, part of his World War I fire-control system.

Bearing rate Rate at which bearing (observed angle of target) changes.

Boat-tailed Tapered at the rear end (of a shell).

Bracket and halve Fire-control technique in which the shooter first places salvoes on either side of the target and then halves the distance between salvoes in the hope of getting a salvo onto the target (if the halved salvoes do not bracket, the shooter adjusts until they do, then resumes halving).

Bracketing The fire-control practice of firing a sequence of salvoes in hopes of placing one to each side (short and over) of the target.

BuOrd Bureau of Ordnance (US Navy).

Burster Explosive filling (shells).

CAFO Confidential Admiralty Fleet Order, a document circulated to the fleet by the Admiralty.

CIC Combat Information Center (US ships).

Coincidence rangefinding Rangefinding by comparing images from two separate lenses, vertical or horizontal halves being brought together and matched. When they coincide, the rangefinder is set for the correct range.

Collimator A device which focuses an image or a spot of light.

Continuous aim Method of firing in which the gun is held on target as the ship rolls, rather than being fired only when the ship rolls so that the cross-hairs are on target.

CPC Common (Armour) Piercing, Capped shell. A capped shell with a bursting charge larger than that of an AP shell.

Consort range Range to a target from a ship acting together with the shooter (the other ship is her consort).

Crh Calibre radius head. A measurement of how pointy a shell head is. The higher the measurement, the pointier the shell.

Cross-cut Technique for deducing target speed and course from a combination of observed range rate and rate across.

Cross-hairs The vertical and horizontal lines in a gunsight or similar device.

Cross-levelling Compensation for tilt across the line of sight or fire, as opposed to levelling, which is compensation for the up and down motion of the line of fire.

Cross-roll Roll across the line of sight, as compared to roll, which is motion of the line of sight up and down. If a gun is pointed directly to the side of a ship, and the ship is simply rolling, the only motion which has to be compensated for is roll. If the same gun is pointed fore and aft, the ship's roll is manifested as cross-roll.

Danger space The distance by which a gun may be mis-ranged, yet will still hit. Definition varied from navy to navy, some taking account of the width as well as the height of the target.

Datum range Range input as a basis for fire-control calculations.

DCT Director Control Tower.

Dead time Time between an observation or calculation and the moment it is used, eg, between when a rangefinder observation is made and when it is received at a calculator.

Deflection rate The speed of the target across the line of fire, ie, the speed needed to lead the target in order to hit. Deflection is measured in knots, but guns are pointed at angles; the angle depends on the range and the deflection.

DGD Director of Gunnery Division (of Admiralty), Royal Navy.

Directorscope Early US term for a director.

Dispersion Spread of where shells land compared to the range to which they are aimed.

DNC Director of Naval Construction, Royal Navy.

DNO Director of Naval Ordnance, Royal Navy.

Dreyer Table An analytic fire-control device invented by Lieutenant (later Vice Admiral Sir) Frederic C Dreyer RN.

DRT Dead Reckoning Tracer.

Dumaresq A device to deduce the range rate and the rate across (deflection) from own and target course and speed and the direction to the target.

Ecartometry (French) measurement of the vector between splash and target.

FCB Fire-Control Box (combined surface and anti-aircraft fire-control system for smaller ships, Royal Navy).

Feedback Comparison between predicted and observed data as a way of checking initial assumptions.

Fire control The entire process of ensuring that a shell will hit a distant moving target, including calculation of target position and compensation for own-ship movement.

FKC Fuse-keeping clocks.

Fleet Problems Major US naval exercises in the interwar period, ending in 1940.

Follow-the pointer Means of ensuring that a device, such as the elevation wheel of a gun, matches a transmitted order, by having its operator move a dial to match a remotely controlled pointer indicating the desired position (as opposed to a visual, a set of numbers displayed by remote control).

GDT Gyro Director Training.

GE General Electric Company (US).

Geometric range The actual distance between shooter and target.

GIS (British) concentration technique based on individual ship control.

GMS (British) concentration technique based on master ship control.

Gun range The range at which a gun is set; because conditions vary, gun range may not be the range the shell actually covers.

Gun trainer/training Mover/movement of a gun in bearing (azimuth).

Gunlayer/gunlaying Mover/movement of a gun in elevation.

HE High explosive (shell).

Heading The direction of a ship's bow.

Hm Hundreds of metres, a common range unit in the Imperial German Navy.

Horizontal spotting Spotting limited to whether a splash was short or over, without any attempt to measure the distance between splash and target.

Hunter Device intended to follow automatically the motion of another; in effect an automated equivalent to the human element of 'follow the pointer'.

IFF Identification Friend or Foe, usually in the context of radar.

Inclinometer Device to measure the angle (inclination) of target to own course.

Integrator Device adding up increments, eg, distances covered at various speeds. Kinematics: description of motions, eg, motions of target and own ship; often distinguished from ballistics, in that the kinematic problem is to determine target course and speed, the ballistic problem being to determine the appropriate gun range to hit that target.

Ladder firing/ranging A ladder is a series of salvoes fired in rapid succession, no attempt at correction being made until the fall of all the salvoes is observed; alternative to *bracketing*, in which two salvoes are fired in rapid succession and their splashes observed, or even single-salvo fire.

Levelling Compensation for the up and down motion of the line of fire.

Line ahead Formation in which ships line up in the direction they are steaming; some alternatives are to steam abreast or to steam so that from overhead the formation seems to form a line at an angle to its direction.

Linear speed Speed in terms of distance, eg feet per second, as distinguished from angular speed, the speed at which something turns or at which the bearing of an object changes. The basis of many integrators is that the linear speeds of positions on a wheel spinning at constant (angular) speed vary depending on how far they are from the centre of the wheel.

Lobe switching Radar technique for precise pointing by comparing a radar image as seen in two beams (lobes) at slight angles to each other.

LOS Line of sight.

Manometer A pressure gauge.

Master ship/master-ship firing Concentration technique in which a master ship controls other ships' fire.

NID Naval Intelligence Department.

ONI Office of Naval Intelligence.

Overs (as in overs and shorts of salvoes) A salvo falling over is falling beyond the target.

Parallax The apparent difference in the position or direction of an object when viewed from different positions, eg, the difference between the images in the viewfinder and the lens of a camera.

Parbuckling Turning shells around on their bases until they were under the hoist.

PIL Position in line.

Pitch Motion of a ship's bow up and down.

Plotting room (US) Below-decks fire-control position, the name being taken from the plots the US Navy used for fire control.

Pointer Gun trainer.

Pointer firing Method of firing in which the pointer was the principal member of the gun crew because he pointed the gun at the target. He fired when the guns seemed to be horizontal (ie, when his cross-hairs crossed the horizon).

Position-keeping Projection forward of the position of the target, so that the position-keeper always knows where the target is; the main function of computers in fire control.

Post à Calcul (PC) French term for the central fire-control position, below decks (literally calculating position).

PPI Plan Position Indicator, a map-like type of radar display.

PZ British pre-1914 term for a tactical exercise, presumably referring to the signal initiating the exercise.

QF (gun) Quick-firing, meaning a gun using metal cartridge cases.

Range rate Rate at which range changes over time (not constant).

Rangefinder control Type of fire control using rangefinder data and minimising spotting; introduced early in World War I.

Rangefinder cut Rangefinder observation, called a cut because in a coincidence rangefinder the image was cut horizontally or vertically, the observer matching the two halves.

Rangefinder range Range as measured by a rangefinder; due to optical distortion, may not be the actual (geometric) range.

Range gate/range gating Radar technique in which the radar concentrates on signals within a gate (ie, within set limits); means of gaining precision and also of excluding clutter.

Range-keeping Projection forward of the range of the target, so that the range-keeper always knows the range (as opposed to position-keeping, which is concerned with both range and bearing).

Range-taking Determination of range, generally referring to a rangefinder.

Rate along Rate at which range changes, actually the component of the range-rate vector in the direction of the target.

Rate across Component of the range-rate vector across the line of sight to the target, proportional to the rate at which target bearing changes; this is the apparent speed of the target.

ROCORD Rate of Change of Range and Deflection indicator (Barr and Stroud Dumaresq equivalent).

Roll Motion of a ship from side to side.

Salvo A group of shells fired together.

Salvo pattern Size of the area in which the shells of a salvo typically fall.

Scartometer Device to measure the vector from splash to target; French term is ecartometer.

Selsyn (short for self-synchronous) Motor designed to follow another device automatically, for remote indication and also to control remote power devices.

Servo control Remote-power control.

Shorts (as in overs and shorts in salvoes) Shells falling at a range short of the target.

Sokutekiban Japanese device to measure target course and speed as an input to fire control.

Spotting Observation of the fall of shot to correct aim.

Spotting correction Correction to fire-control solution based on observation (spots).

Spotting glass In the US Navy, a device used by a spotter to get a better idea of the positions of splashes and targets; it consisted of two widely separated lenses, each feeding an eyepiece for one eye, so that the spotter had a stereo view.

Stable vertical (or stable zenith) A device which defines a vertical direction even though a ship rolls and pitches, generally based on gyros.

Stadimeter Rangefinder based on measuring the angle between two parts of an object separated vertically, eg, waterline and masthead.

Stereoscopic rangefinding Rangefinding by matching the separate images seen by the two eyes of the range-taker in the observer's brain.

Straddles Fall of shot in which shells of a salvo fall on either side of a target (generally those falling on the target itself are not as visible as the splashes); the goal of fire control is to cause straddles.

Synchro Any device which automatically follows another; a selsyn is a type of synchro.

Super-elevation Additional elevation to correct for individual gun conditions or positions, corresponding to desired gun range.

Synthetic position-keeping Position-keeping based on an assumed target course and speed, calculation being compared with later observation as a means of feed-back.

Tachymetric Method based on observed target speed, usually in angular terms.

TEM Towed Electrode Method.

Throw-off firing (also, off-set firing) Method of firing practice in which the shooter fires at a manoeuvering ship, but the aim point is offset so that he does not hit it; a means of learning to fire at the most realistic possible moving target.

TIC Time Interval Compensation.

Transmission interval The estimated dead time between computing and firing.

Transmitting station (GB) Central computing station of British fire-control systems.

Trunnion tilt Tilt of guns across the line of fire.

UEP Underwater Electric Potential.

Variable-speed drive Device converting variable position into variable speed for integration.

Vertical spotting US Navy technique attempting to measure the distance between splash and target by the apparent difference in vertical positions as seen from a masthead.

Vickers Clock Device giving range at a particular time on the basis of a range rate; the first major fire-control calculator.

Yaw Movement of the head of a ship from side to side, away from the ordered course, due to the interaction of roll and rudder in waves (some modern ships use the inverse effect, rudder action to counteract roll).

Bibliography

Primary Sources

For archival and private sources used, please refer to the footnotes for each chapter.

Secondary Sources

Apal'kov, Yu V, *Udarniye Korabliy* (St. Petersburg: Galeya Print, 2003).

Bagnasco, Erminio, *Le Armi Delle Navi Italiane Nella Seconda Guerra Mondiale* (Parma: Ermanno Albertelli, 1978).

Bargoni, Franco and Gay, Franco; and Ando, Elio, *Le Corazzate Italiane nella Second Guerra Mondiale* (Rome: Edizioni Bizzari, n.d. [probably late 1970s]).

Brooks, John, *Dreadnought Gunnery and the Battle of Jutland: The Question of Fire Control* (Abingdon: Routledge, 2005).

Brown, D K, *The Grand Fleet: Warship Design and Development 1906–22* (London: Chatham, 1999).

Brown, G I, *The Big Bang, A History of Explosives* (London: Sutton, 2005, paperback reprint of 1998 edition).

Burt, R A, *British Battleships 1889–1904* (London: Arms and Armour, 1988).

———, *British Battleships of World War I* (London: Arms and Armour, 1986).

———, *British Battleships 1919–39* (London: Arms and Armour, 1993).

Campbell, John, *Naval Weapons of World War Two* (London: Conway, 1985).

———, *Jutland: An Analysis of the Fighting* (London: Conway Maritime Press, 1986).

Chiabotti, Stephen D ed., *Tooling For War: Military Transformation in the Industrial Age* (Chicago: Imprint Publications, 1996).

Corbett, Julian S, *Maritime Operations in the Russo-Japanese War 1904–5* (2 vols; Annapolis: Naval Institute Press, 1994, reprint of edition published by Admiralty Naval Intelligence Division in 1914 and 1915, with introduction by John B Hattendorf and Donald M Schurman).

Custance, Admiral Sir Reginald, *The Ship of the Line in Battle* (London: William Blackwood, 1912).

Douglas, General Sir Howard, *A Treatise on Naval Gunnery 1855* (reprint of fourth edition by Conway Maritime Press, 1982).

Dumas, Robert, *Les Cuirasses Dunkerque-Strasbourg, Richelieu, Jean Bart* (Bourg en Bresse: Marines, 2001).

Dumas, Robert and Guiglini, Jean, *Les Cuirasses de 23,500 tonnes* (Outreau: Lela Press, 2005).

Evans, David C and Peattie, Mark R, *Kaigun: Strategy, Tactics, and Technology in the Imperial Japanese Navy, 1887–1941* (Annapolis: Naval Institute Press, 1997).

Friedman, Norman, *US Naval Weapons: Every Gun, Missile, Mine and Torpedo Used by the US Navy from 1883 to the Present Day* (Annapolis: Naval Institute Press, 1983).

———, *Naval Radar* (London: Conway, 1981).

Fioravanzo, Amm. di Sq. (C.a.) Giuseppe, *L'Organizzazione della Marina durante il conflitto*, Vol 1, *Efficienza all'apertura della ostilita* (Vol XXI of the Italian official WW II naval history: Rome: Ufficio Storico della Marine Militare, 1972).

Fiske, Rear Admiral Bradley A, *From Midshipman to Rear Admiral* (New York: Century, 1919).

Forstmeier, Friedrich and Breyer, Siegfried, *Deutsche Grosskampfschiffe 1915–18* (Munich: Lehmann, 1970).

Garier, Gerard and du Cheyron, Patrick, *Les Croiseurs Lourds Francais Duquesne et Tourville* (Outreau: Lela Press, 2003).

Gay, Franco, and Gay, Valerio, *The Cruiser Bartolomeo Colleoni* (London: Conway Maritime Press [Anatomy of the Ship Series], 1987).

Gordon, Andrew, *The Rules of the Game: Jutland and British Naval Command* (London: John Murray, 1996).

Green, Jack, and Massignani, Alessandro, *The Naval War in the Mediterranean 1940–3* (London: Chatham, 1998).

Guiglini, Jean and Moreau, Albert, *Les Croiseurs de 8000 tonnes* (Bourg en Bresse: Marines, n.d.).

Hase, Commander Georg von, *Kiel to Jutland* (New York: E P Dutton, 1922).

Hodges, Peter, *The Big Gun: Battleship Main Armament 1860–1945* (London: Conway Maritime Press, 1981).

Jellicoe, Admiral Sir John *The Grand Fleet 1914–16: Its Creation, Development, and Work* (originally published 1919; new edition published 2006 by Ad Hoc Books, London, with foreword by Dr Eric Grove).

Kemp, Lieutenant Commander P K, *The Papers of Admiral Sir John Fisher* (Vol I, London: Navy Records Society, 1960).

Kingsley, F A, ed, *Radar and Other Electronic Systems in the Royal Navy in World War II* (London: Macmillan, 1995).

Koop, Gerhard and Schmolke, Klaus-Peter, *Die Linienschiffe der Bayern-Klasse* (Bonn: Bernard & Graefe, 1996).

———, *Von der Nassau-zur Koenig Klasse* (Bonn: Bernard & Graefe, 1999).

Lacroix, Eric, and Wells, Linton, *Japanese Cruisers of the Pacific War* (Annapolis: Naval Institute, 1997).

Lassaque, Jean, *Le Croiseur Emile Bertin 1933–59* (Bourg en Bresse: Marines, n.d., probably 1993).

———, *Les C.T. de 2880 Tonnes du Type Mogador 1936–45* (Bourg en Bresse: Marines, 1996).

———, *Les CT de 2800 Tonnes du Type Fantasque* (Nantes: Marines, 1998).

———, *Les CT de 2700 Tonnes du Type Vauquelin* (Bourg en Bresse: Marines, 2000).

———, *Les Contre-Torpilleurs Epervier et Milan 1931–46* (Bourg en Bresse: Marines, n.d.).

———, *Les CT de 2400 Tonnes du Type Jaguar* (Bourg en Bresse: Marines, n.d.).

Legemaate, H J, Mulder, A J J, and van Zeeland, M G J, *Hr. Ms. Kruiser "De Ruyter" 1933–42* (Purmurend [Netherlands]: Asia Maior, 1999).

———, *British Cruisers of World War II* (London: Arms and Armour, 1980).

Mallett, Robert, *The Italian Navy and Fascist Expansionism1935–40* (London: Frank Cass, 1998).

Marder, Arthur J, *Jutland and After, May 1916 – December 1916* (2nd ed of Vol 3 of *From the Dreadnought to Scapa Flow*; Oxford, 1978).

———, ed, *Fear God and Dread Nought* (Fisher correspondence collection: 3 vols, all London: Cape; 1952, 1956, 1959).

McLaughlin, Stephen, *Russian and Soviet Battleships* (Annapolis: Naval Institute Press, 2003).

Mindell, David A, *Between Human and Machine* (Baltimore: Johns Hopkins, 2002).

Morrison, Rear Admiral Samuel E, *The Struggle for Guadalcanal, August 1942–February 1943* (Vol 5 of the *History of United States Naval Operations in World War II*; Boston: Little, Brown, 1949, reprint 1975).

———, *Leyte, June 1944–January 1945* (Vol 12 of the *History of United States Naval Operations in World War II*; Boston: Little, Brown, 1958, reprint 1974).

Moss, Michael and Russell, Iain, *Range and Vision: The First Hundred Years of Barr & Stroud* (Edinburgh: Mainstream, 1988).

Moulin, Jean, *Les Croiseurs de 7600 Tonnes* (Bourg en Bresse: Marines, 1993).

Moulin, Jean and Maurand, Patric, *Le Croiseur Algerie* (Bourg en Bresse: Marines, n.d., probably 2002).

Mullenheim-Rechberg, Burckhard Baron von (transl. Jack Sweetman) *Battleship Bismarck: A Survivor's Story* (Second Edition) (Annapolis: Naval Institute Press, 1990).

Parke Hughes, Thomas, *Elmer Sperry: Inventor and Engineer* (Baltimore: Johns Hopkins Press, 1971).

Patterson, A Temple, *The Jellicoe Papers* Vol I (Navy Records Society, 1964).

Peira, M P, *Historique de la Conduite du Tir Dans La Marine* (3 vols; Memorial de L'Artillerie Francaise, 1955).

Pollen, Anthony, *The Great Gunnery Scandal: The Mystery of Jutland* (London: Collins, 1980).

Prados, John, *Combined Fleet Decoded: The Secret History of American Intelligence and the Japanese Navy in World War II* (Annapolis: Naval Institute Press, 2001).

Ranft, B McL, ed, *The Beatty Papers* (London: Navy Records Society, 2 vols, 1989 and 1993).

Raven, Alan, and Roberts, John, *British Battleships of World War II* (London: Arms and Armour, 1976).

Roskill, Stephen, *Naval Policy Between the Wars: Vol. 2, The Period of Reluctant Rearmament, 1930–9* (London: Collins, 1976).

Salou, Charles, *Les Torpilleurs d'Escadre du Type Le Hardi 1932–43* (Outreau: Lela Presse, 2001).

Schmalenbach, Paul, *Die Geschichte der deutschen Shiffsartillerie* (Herford; Koehlers, 1968).

Shirokorad, Aleksandr, *Oruzh'ye Otechestveennogo Flota 1945-2000* [Domestic Naval Weapons] (Moscow: Kharvest, 2001).

Sumida, Jon Tetsuro, *In Defense of Naval Supremacy: Finance, Technology, and British Naval Policy 1889–1914* (Boston: Unwin Hyman, 1989).

———, ed, *The Pollen Papers 1901–16* (London: Navy Records Society, 1984).

Taylor, Bruce, *The Battlecruiser Hood: An Illustrated Biography 1916–41* (London: Chatham Books, 2005).

Whitley, M J, *German Cruisers of World War II* (London: Arms and Armour, 1985).

Index

Figures in italic refer to references in captions those in bold italic to the main subject of the illustrations

A

Abruzzi (Italy), 267
Achilles (UK), 129 155
Admiral Scheer (Germany), *169*, *170*
Admiralty Fire-Control Clock (AFCC), 129–30, 134, *134*, *135*, 186, 230
Admiralty Fire-Control Table (AFCT), *14*, 42, *43*, *45*, *51*, *113*, 117–19, *121*, 123–5, *126*, 129–30, *135*, *136*, *137*, *139*, 145, 146–7, ***153–4***, 184, 224, 232
Admiralty Research Laboratory (ARL), 15, *39*, 129
Africa (UK), 60, 74, 76
Agamemnon (UK), 78
Agincourt (UK), 46, 80, 98
Aichi Clock Company, 229
Aigle class (France), *257*
air-observation, 102, *110*, 111, 119, 138, 149, 153, 155, 189, 191, 225, 236, 239, 258, 279
Ajax (UK), 24, 80, 129, 155
Ajax class (UK), *126*
Aki (Japan), 233
Alabama (USA), 176, ***290***
Aland Islands, Battle of, 278
Alaska class (USA), *64*, *210*, 215, ***290***
Alberico Da Barbiano (Italy), ***261***
Aleksandr Suvorov (USSR), ***276***
Aleutian Islands, Battle of, 219
Algérie (France), *247*
Almirante Brown (Argentina), ***263***
Almirante Brown class (Argentina), 263
Almirante Latorre (Chile), *75*, 133, ***228***
Almirante Latorre class (Chile), 65
Amazon (UK) 132, 134
Ambuscade (UK), 132
Andrea Doria (Italy), 262, ***260***, ***267***
Anson, Lieutenant, 23
Anson (UK), ***150***
anti-aircraft
 computers, *136*, 168
 directors, *39*, 201–2
 fire, 27
 radar, 59
Aoba class (Japan), *239*, *240*
Arethusa (UK), 23
Argo Company, 58, 62
Argo Clock, *54*, 60, 62, 72, 79, 108, 123, 248, 274
 see also Pollen Clock
Ark Royal (UK), 149
Ark Royal class (UK), 151
Arkansas (USA), *28*, 179, 182, *192*, *195*, 205
Arkansas class (USA), *290*
Arizona (USA), *196*, 203, 285
Arma system, *196*, 205
armour, *10*, 115–17, *230*, 235, 287–90
Armstrong, 285
Asahi (Japan), 36–7
Astoria (USA), ***202***
Atlanta (USA), 216
Attilio Regolo class (Italy), *124*
Auckland (UK), 155
Audacious (UK), 96
Aurora (UK), *113*
Australia (Australia), 80, ***107***, ***127***, ***128***

B

Bacon, Admiral Reginald, *32*, 43, 57, *70*, 76, 85, *281*
Baden (Germany), 111, ***161***
Baden class (Germany), 159, *161*
Barents Sea, Battle of, 154, 155
Barham (UK), *25*, 115, ***116***, 117, *118*, 129, 147, 153, 258
Barr & Stroud, 23, 28, 31, 33, 61, 130–3, 134, 160, *167*, 169, 178, 229, 230, 245, 272

 see also fire-control systems; rangefinders; ROCORD
Bartolomeo Colleoni (Italy), 151
Battenberg, Admiral, 85–6
'Battle' class (UK), *151*
Bausch & Lomb, 178, 203
Bayern (Germany), 158
Béarn (France), *56*, 253
Beatty, Admiral David, 87, 93, 94, 98–9, 108–9
Belfast (UK), *126*, 144
Bellerophon (UK), 33, 76, 78, ***104***
Bellerophon class (UK), *70*
Benbow (UK), 106
Bergamini, Capitano di Corvetta Carlo, 262
Berwick (UK), ***122***, ***128***
Bismarck (Germany), *14*, *39*, 65, 111, 119, *137*, 142, 143, 148, 149–50, 154, 155, *173*, 174, *174*
Bismarck class (Germany), 169, *171*, *172*
Black Prince (UK), 166
Black Swan class (UK), 135
Blücher (Germany), 91, 108, 159
Boer War, 43, 289
Bofors, 168
Boise (USA), 235, 241, 285
Bonaventure (UK), 142
Borea (Italy), 263
Boucheron, Lieutenant de Vaisseau du, 247–8
Bouvet (France), 244
Brassey's Naval Annual, 20–1, 22, 64
Breguet, 253
Bretagne (France), 154, 252
Bretagne class (France), 248, 251, 252, 258
Bridgman, Vice Admiral Sir Francis, *32*
Brisbane (UK), 53
Britannia (UK), *21*
Brooklyn (USA), ***202***
Brooklyn class (USA), *64*, *149*, 200, *204*, 209
Brooks, John, *32*
Brown, D K, 35
Brownrigg, Commander J, 47
Bureau of Ordnance (BuOrd, USA), *8*, 15, 179, *180*, 182, 186, 187, *189*, *196*, *198*, 199, 200–3, *219*, *223*, 260, *290*

C

C class (UK), 46, *52*, 134
Caesar (UK), ***16–17***
Caio Duilio (Italy), 150
Cairo (UK), 52
Calabria, Battle of, 150, 153
Caledon (UK), 53
California (USA), *178*, *185*, ***188***, *216*, 221, 223, 224
California class (USA), *64*, 178, 179, *184*, 186, *188*, 190–1, *192*, 203, 216
Callaghan, Rear Admiral Daniel J, 216
Callaghan, Admiral Sir George A, 87, 91–2, 93, 96, 97
Calliope (UK), 53
Calypso (UK), 53
Cambrian (UK), 53
Canada (UK), *75*, 102, *228*
Canberra (Australia), *128*, 136
Canterbury (UK), 53
Cape Esperance, Battle of, 215, 241
Cape Matapan, Battle of, 114, 115, 155, 267
Cape Sarych, Battle of, 273, 277
Cape Spartivento, Battle of, 155
Capetown (UK), 52
Cardiff (UK), *53*, 104
Carlisle (UK), 52
Carysfort (UK), 53

Castle, Lieutenant Commander, 186
Castor (UK), 53
Cavour class (Italy), *263*, *264*
Centaur (UK), 53
Centurion (UK), 80
Ceres (UK), 53
Chack, Lieutenant de Vaisseau Paul, 248
Champion (UK), 53
Chapaev (USSR), ***275***
Chapaev class (USSR), *273*, *275*, *277*, 279
Chester (USA), ***198***, ***201***
Chokai (Japan), ***240***
Churchill, Winston, *35*, 62, 85, 90
Clausen, Lieutenant Hugh, 49, 62, 123
Cleopatra (UK), 53
Cleveland class (USA), *64*, *149*, 224
Cochrane (UK), 81
code-breaking, 107, 209, 238
Colbert (France), *257*
'Colony' class (UK), 130
Colorado (USA), *192*, ***215***, ***218***
Colorado class (USA), *113*, 188, 190–1, *192*, 239
Colossus (UK), 74, 88, ***88–9***, 105
Colossus class (UK), *32*, *281*
Columbo (UK), 52
Combat Information Center (CIC), 215, 223, 240
compasses
 gyro, 20, *43*, 44, 45, 178, 183, 229, 251, 254, 279
 magnetic, 43
Comus (UK), 53
Concord (UK), 53
Connecticut class (USA), 179
Conqueror (UK), 80
Conquest (UK), 53
Constance (UK), 53
Conte di Cavour (Italy), 260, 262, ***265***
continuous aim, 19, 20, 89, 176, 188–9, 245, 247, 262
Corbett, Julian, 107
Cornwall (UK), *128*
'County' class (UK), *240*
Courageous class (UK), 48, *102*
Courbet (France), 241
Courbet class (France), *258*
Coventry (UK), 53
cross-levelling, 18, *113*, 133, 151, *169*, 192, *199*, *205*, *223*, *247*
cross-roll, *19*
Curacao (UK), 53
Curlew (UK), 53

D

D class (UK), 46, *52*
Dakar, 258
danger space, 18, 37, 69, 73, 86, 88, 154, 158, 216, 234
Dante Alighieri (Italy), 262, *271*, ***284–5***
Danton class (France), 244, 248, 251
Dartmouth (UK), 53
data transmission, 23, 31–3, 35–7, *132*, 160, 164, 182, 211–13, 244, 246, 252, 258, 262, 266, 270, 273–4, 278
 see also transmitting station
Davydov (Russian officer), 270
De Feo, Capitano di Fregata Vincenzo, 262
De Grasse class (France), *255*
De Ruyter (Holland), 169
De Zeven Provincien class (Holland), 284
Defence (UK), 166
Delaware (USA), *177*, *178*, 179, ***286***, 287
Delaware class (USA), *188*
Delhi (UK), 52

Denver (USA), ***219***
Derfflinger (Germany), 108, 109, *159*, 160, 161, 163, 166
Des Moines class (USA), 200, *215*, 224, 225
Detroit (USA), *194*
Deutschland (Germany), *168*, 169, *202*
Dido (UK), 43, ***131***
Dido class (UK), *113*, 142, 147
Direction d'Artillerie Navale, 253
Director Control Tower (DCT), *13*, *14*, 56, *113*, 117, ***120***, 123, ***124***, 126, ***128***, ***129***, 130, ***131***, ***134***, ***138***, 144–7, 148–50, *247*, 250, 264, 266–7, *269*, 278
Director of Naval Construction (DNC) (UK), *14*, 74, 98, 99
Director of Naval Ordnance (DNO) (UK), 12–14, 20, 21, 24, *32*, 35, 36, 38–9, 43, 54, 58, 61, 74, 75–6, 78, 90, 117, 118, 123, *281*, 289
directors, 73–81, *92*, *94*, 104–5, 106, *113*, 130, 133, 161, 168, 171–2, 179–80, *184*, 228, ***229***, 231–2
 AA (British), *234*
 AA (Japanese), ***236***, ***237***
 AA (Soviet), *275*
 Argentinian, *263*
 Australian, ***107***, ***127***
 British, *109*
 French, *247*, *249*, *251*, 252, *253*, *255*, *258*
 Italian, ***261***, *262*, ***264***, ***265***, ***266***, ***267***
 Mk VI (USA), *188*
 Mk VII (USA), ***188***, *190*
 Mk VIII (USA), *190*, *193*
 Mk IX (USA), *190*, *205*
 Mk X (USA), *190*, *190*
 Mk XI (USA), *13*, *193*
 Mk XIII (USA), *190*, *205*
 Mk XVI (USA), *193*
 Mk XVII (USA), *193*
 Mk XVIII (USA), *193*
 Mk XX (USA), *13*, ***195***, *196*
 Mk XXI (USA), *195*
 Mk XXIV (USA), *190*
 Mk 2 (USA), *28*
 Mk 4 (USA), *28*
 Mk 7 (USA), *28*
 Mk 9 (USA), *188*, *190*, *205*
 Mk 18 (USA), *191*
 Mk 19 (USA), *13*, *201*, *234*, *223*
 Mk 20 (USA), ***196***, *217*
 Mk 22 (USA), *191*, ***194***
 Mk 23 (USA), *194*
 Mk 24 (USA), *194*, *196*, *198*, *199*, *200*, *202*
 Mk 27 (USA), ***199***, *223*
 Mk 29 (USA), *190*, *199*
 Mk 31 (USA), ***197***, ***202***, *205*, *225*
 Mk 32 (USA), *190*, *205*
 Mk 33 (USA), ***198***, *237*, ***204***, *223*, *225*, *226*
 Mk 34 (USA), ***202***, ***204***, *218*, *219*, *220*, *221*, ***222***, *224*, *226*
 Mk 35 (USA), ***226***
 Mk 37 (USA), *191*, *201*, ***212***, *224*
 Mk 38 (USA), ***208***, ***209***, ***210***, *212*, *226*
 Mk 41 (USA), *205*
 Mk 43 (USA), *205*
 Mk 55 (USA), *212*
 Russian, 270, 274
 Scott's (UK), *73*
 Type 14 (Japan), *239*
 Type 91 (Japan), *234*
 Type 94 (Japan), *231*, *236*, *237*, *239*
 Type 98 (Japan), *231*, ***237***
 Soviet, *273*
divisional tactics, 86–7
Dominion (UK), 81

INDEX

Dogger Bank, Battle of, 90, 93, 94, 96, 98, 104, 108–10, 159, *159*, *283*, 284, 285
Dorsetshire class (UK), *113*
Douglas, Sir Howard, 22
Dove, Lieutenant S, 49, 123
Down, Commander R T, 53, 186
Downes (USA), 186
Drax, Admiral Lord, 59
Dreadnought (UK), *21*, *32*, 33, 57, **70–1**, 74, 76, *104*, *281*
Dreyer, Vice Admiral Frederic C, 42, 43–4, *43*, 46, 57, 80, 250
Dreyer, Captain J T, 42
Dreyer calculator, *48*, *49*, 110, 121, 228, 250
Dreyer Table, 11, 29, 30, *30*, 41, *41*, 42, 45, *45*, 46, 47–53, *48*, *49*, **50–2**, 60–1, 62, 65, 71–2, 89, 92, 93, 104, 108, 119, 121–3, *126*, 133, 139, 154, 177, 184, 228, 248, 250, 251
Drottning Victoria (Sweden), *167*
Duca d'Aosta class (Italy), *261*, *264*, *266*
Duffy, Captain A A M, *32*
Duguay-Trouin class (France), *247*
Duke of Edinburgh (UK), 35–6
Duke of York, 142, 144, 145, 146, 154
Duilio (Italy), 262, 264
Duilio class (Italy), *263*, *266*
Dumaresque, Rear Admiral John Saumarez, 29, 30, 43
Dumaresques, 29–31, *29*, *30*, *31*, *33*, 41, 44, 45–7, 48–9, 56, 59, 68–9, 71, 96, 111, 119, 121, 122, 182, 228, 245–6, 250, 260
see also rangefinding
Duncan (USA), 215
Dunedin (UK), *52*, **109**
Dunkerque (France), *249*, *251*, 256–7
Dunkerque class (France), 244, 255, 256, 257, *264*, *285*
Duquesne class (France), 254
Durban (UK), *52*

E

E class (UK), 46
electric logs, 28
electronic identification, 213
Electropribor, 278
Eller, Lieutenant Commander E M, *8*
Elliott Brothers, *45*, 46, 111, 123
Elphinstone, Keith, 46, 48, 123
Emden (Germany), 167
Emerald (UK), *125*
Emile Bertin (France), *247*
Emperor of India (UK), 49, 52, **280–1**
Empress Augusta Bay, Battle of, 219
Empress of India (UK), 80, 115–17, 186
Encounter (UK), 241
Eng transmission system, 244
Engineering (magazine), 21
Enterprise (UK), 123, **124**, *129*, 130
Erie class (USA), 203, **226**
Erin (UK), 80
Ersatz Monarch class (Austria-Hungary), *165*
Evan-Thomas, Captain, 76, 78
Evershed & Vignolles, 33, 111
Evertsen class (Holland), 168
Evstafii (Russia), 273
Excellent (RN gunnery school), 15, 20, 22, 59, 69, 74, 80, *126*, *131*, *135*, 142
Exeter (UK), *126*, *127*, *129*, *129*, 241

F

Falkland Islands, Battle of, 98, 106–7
Fantasque class (France), 97, *257*, 258
Fargo class (USA), *212*, 224
Fiji class (UK), *113*, *124*, *128*, *131*, 149

fire-control computers, *11*, *29*, 36
see also Dreyer Table; Pollen Clock
fire-control systems
 American, 176–9, 182–93, **187**, **189**, **193**, 199–203
 analytic, 47–53, 59, 167
 Barr & Stroud, 130–3, 263
 Bureau-Sperry, 186, **189**, **191**, 262
 Dutch, 167–9
 Erikson, 274–5
 'Eversheds', 33
 Fire Control Box (FCB), 135
 French, 245–6, 247–57
 Fuse-Keeping Clock (FKC), 134
 Geisler, 270, 274, 275, 277, 278
 Italian, 260–7, **262**
 Japanese, 228–33
 and radar, *28*, 142, 147–8
 Russian, 39, 270–1, *271*, 272, 274–5, 277
 synthetic, 42, 53–61, 123, 133, *135*, 148, 167, *174*, 228, 252, 278
 trials (UK), 33–5, 78–80
 Soviet, 269, 278–9
fire-control tops, *21*, 24, *32*, *36*
Fisher, Admiral Sir John Arbuthnot, 20, 22, 55–6, 288
Fiske, Rear Admiral Bradley, 23
Florida, 24, *175*, *178*, 179
Ford, Hannibal C, *56*, 182
Ford integrator, *56*
Ford Range-keeper, *see* range-keeping
France (France), *38*
Francesco Nullo (Italy), 150
Freccia class (Italy), 266
friendly-fire, 84, 106, 213, 216, 222
Frunze (USSR), 275
Fubuki class (Japan), 237
Furious (UK), *110*
Furious class (UK), *102*
Furutaka class (Japan), *239*, *240*
Fuso (Japan), 222
Fuso class (Japan), 230, *234*, 241

G

Gallipoli campaign, *101*
Gangut (Russia), 274
Gangut class (Russia), *271*, 274, 275
Garibaldi class (Italy), *261*
Geisler and Company, 270, 274, 275, 277, 278
General Crauford (UK), 49
General Electric (GE), 164, 187, *188*, 191–2, 199
General Wolfe (UK), *110*
Geneva Conference (1927), 115
geometric range, *see* range
George Leygues (France), 258
Germain transmission system, 244–5
gimetro (bearing gyroscope), 262–3
Girardelli, *167*, 187, 252, *260*, 262, 263
Giulio Cesare (Italy), 262
Glasgow (UK), **138**
Glatton (UK), *52*
Gloucester (UK), 153, 267
Gneisenau (battlecruiser, Germany), 150, *171*
Gneisenau (cruiser, Germany), 106
Goeben (Germany), 166, 273, 277–8
Good Hope (UK), 75, 76, 78, 106
Goodall, Stanley V, *183*, 288
Goodenough, Lieutenant William, 43
Gorgon (UK), *52*
Graf Spee (Germany), *39*, **170**
Greyhound (UK), 155
Guadalcanal, Battle of, 154, *207*, 209, 210, 211, 213, 216, 219–21, *219*, *235*, 241
Guépard class (France), *257*
Gunnery Manual (publication), 93, 96

Gustav V (Sweden), *167*
gyro-controlled torpedoes, 22
Gyro Director Training (GDT), 49–52, 105, *110*
gyro-firing mechanism (von Petravic), 164, 166, 252
gyro room, 123
gyro-stabilised rangefinders, *23*, 28, 30, 54, 58, 80

H

Haguro (Japan), 241
Haida (Canada), *135*
Harding, Captain Edward W, 21, 28
Hartlepool, 107
Hase, Captain Georg von, 166
Hawkins class (UK), *127*
Hazemeyer, 132, 167–8, *167*, 263
Helena (USA), *202*, 211, 213, 216–19
Heligoland Bight, Battle of, 94, 106, 109
Henderson gyro, 105
Henley, J C W, 61
Hercules (UK), 86, *89*
Hero (UK), 35, 74
Hibernia (UK), *21*, 81
Hiei (Japan), 216, 233
Hindenburg (Germany), **162–3**
Hipper class (Germany), 142, 169, *172*
Hoche (France), 248
Holland, John, 54
Holland, Admiral Lancelot, 154
Hood (UK), *8*, 48, 49, *50*, 53, 65, 111, 115, 117, 119, 121, *137*, 142, 143, 146–7, 149, 154, *172*, 174, *174*
Hood class (UK), 99
Hopewell (USA), 188
Howe (UK), **14**
Hughes-Onslow, Captain Constantine, 58
Hussar (UK), 72

I

Iachino, Capitano di Corvetta Angelo, 262
Idaho (USA), **183**, **197**
Idaho class (USA), 178, 180, 202, *202*
Iéna (France), 285
Illustrious (UK), *17*
Imperator Alexandr II (Russia), 270
Imperator Nikolai I (Russia), 274
Imperator Pavel I class (Russia), 274
Imperatritsa Ekaterina Velikaia (Russia), 277–8
Imperatritsa Maria (Russia), 285
inclinometers, *43*, 111, 130, 132, 148, 201, 231, 265, 266
 see also sokutekiban
Indefatigable (UK), *32*, 89, 161
Indianapolis (USA), 199, 203, **223**
inertial guidance systems, 20
Inflexible (UK), 77, 106, **106**
Ingenohl, Admiral von, 158
integrator, *see* position keeping
Invincible (UK), **10**, 98, 106
Ioann Zlatoust (Russia), 273
Iowa (USA), 10, **66–7**
Iowa class (USA), 64, 203, *210*, *212*, *215*, *235*
Iron Duke (UK), *92*, 93, **94–5**
Iron Duke class (UK), 46, 80, 85, *92*
Ise class (Japan), *241*
Isherwood, Lieutenant Commander Harold, 42, 123, 184

J

Jackson, Captain Thomas, 36, *70*
Jaguar class (France), *257*
Jamaica (UK), 155

jamming, 144
Janus (UK), 153
Jauréguiberry (France), 248
Java (Holland), 168
Java Sea, 241
Jean Bart (France), *243*, *253*
Jean Bart class (France), 251
Jellicoe, Rear Admiral John, 14–15, 20, 54, *70*, 74, 75–6, 78–9, 80–1, 84, 85, 86, 87–8, 89, 90, 91–3, 94, 96, 98–9, 106, 107–8, 109
Jupiter (UK), 54, 75
Jutland, Battle of, 25, 86, 87, 90, 93, *94*, 96, 98–9, *98*, 105, 108, 110–11, 113, 119, 159–60, *159*, 161, 163, 166, 249, 284, 285

K

'K' class (Germany), *168*
Kaiser (Germany), 90–1, 111, *283*
Kawachi (Japan), 285
Kelvin, Lord, 42
Kennish, Carpenter William, 72
Kent (UK), *128*
Kent class (UK), *113*, *122*, *126*, 129, 130, 139
Kenya (UK), *131*
Kimberley (UK), 150
King Edward VII (UK), 74
King George V (UK), *32*, 142, 144, 155
King George V class (UK), 24, 35, *52*, 60–1, 69, 80, 87, *113*, 115, 117, *118*, *124*, 130, *131*, *137*, 150
Kirishima (Japan), *207*, *208*, 219–20, 221
Kirov (USSR), **273**
Kirov class (USSR), *273*, *277*, 278
Klado, N L, 271
Knox, S G, 262
Koenig class (Germany), *162*
Kolombangara, Battle of, 219
Komandorski Islands, Battle of, *43*, 225–6, 241
Kongo (Japan), 132, 228, *228*, 229, 233
Kongo class (Japan), 230, *235*, 237, *241*
Krasnyi Kavkaz (USSR), 278
Kronstadt class (USSR), 279
Kula Gulf, Battle of, 216–19
Kure Arsenal, 230

L

La Argentina (Argentina), 133
La Galissionière class (France), 255
Lafrogne, Lieutenant de Vaisseau, 245–6
Lamotte-Picquet (France), *247*
Lamotte-Picquet class (France), *247*
Landstad, D H, 42, 123
Langsdorff, Captain von, *39*
Latouche-Tréville (France), 244
Le Fantasque (France), 258
Le Hardi class (France), *257*
Le Prieur, Ensign de Vaisseau Yves, 248, 250, 251, 252, 253
Le Prieur tables, 248, 250, 251, 252, 253, *260*
Le Triomphant (France), *257*
Leander (UK), 129, 155
Leander class (UK), 129
Leary, Captain H F, 202–3
Lecomte, Lieutenant de Vaisseau, 246
Lee, Commander Willis A, Jr, 201–2
Leonardo da Vinci (Italy), 285
Leipzig (Germany), 169
Leningrad class (USSR), 278
levelling, 18; *see also* cross-levelling
Lexington class (USA), 180, 191, *193*
Leyte Gulf, 221, 232
Liberté (France), 285

Linotype Company, 42
Lion (UK), *32–3*, 35, 98, 108, *108*, 109
Lion class (UK), *32*, 35, *36*, 60, 79, *113*, 130, *137*
Littorio class (Italy), *261*, *264*, *266*
Liuzhol' (Russian officer), 270
Liverpool (UK), *131*, 153
Lloyd, Lieutenant, 23
London (UK), *128*
London class (UK), *113*, *128*, 130, 139
London Treaty, 113, *149*, 264
Lord Clive, (UK), 49, *110*
Lord Nelson (UK), 60
Lorraine (France), 252
Louisiana class (USA), *178*
Louisville (USA), *194*, *196*, **197**
Luigi di Savoia duca degli Abruzzi class (Italy), *264*
Lutzow (Germany), 111, *170*

M

McKenna, Reginald, 78
Mackensen class (Germany), 159, *161*, *162*
Madagascar, 38
magazines, 63, 97, 98–9, *98*, 111, 113, 115, 117, 119, *159*, 282–5, *288*
Magnificent (UK), *17*
Magslip, 129
Maine (USA), 285
Majestic (UK), *17*
Majestic class (UK), *77*, *100–1*
Malaya (UK), *116*, 150
Marat (USSR), *278*
Markgraf (Germany), 166
Marlborough (UK), 80, 96, 117
Mars (Germany), 158
Marshall Islands, Battle of, 223
Maryland (USA), *215*, 221, 223
Maryland class (USA), 164, 178, *184*, *185*, 187, *190*, *193*, *205*, *290*
Matapan, *see* Cape Matapan, Battle of
Mauritius (UK), *131*
Maxim Gorkiy class (USSR), *279*
May, Admiral Sir William H, 44–5, 86
Maya (Japan), 225, *240*
mechanised fire control, *see* fire-control computers
Melbourne (UK), *53*
Merrill, Rear Admiral Stanton A, 219
Mers-el-Kebir, 154
Miakishev (Russian officer), 270
Michigan (USA), *175*, 179
Mikasa (Japan), 287
Minas Gerais (Brazil), 186
Minneapolis (USA), *197*, 216
Minotaur (UK), 81
Minotaur class (UK), *147*
Mississippi (USA), *65–6*, *183*, *188*, *197*, 221–2, 223, *288*
Missouri (USA), *210*
M-motor, 125–6
Mogador class (France), *257*
Mogami class (Japan), *127*, 231, 236, *239*, *241*
Moltke (Germany), *283*
Moltke class (Germany), *157*
Monarch (UK), 46, 61, 80
Montcalm (France), *255*, 258
Moore, Captain Archibald, 76, 78
Mouton, Rear Admiral W A, 61, 132, 169
Mutsu (Japan), **228**, 285
Myoko (Japan), *239*, 240

N

Nachi (Japan), 225, *241*
Nachi class (Japan), *239*, *240*
Nagato (Japan), **227**, **228**, **229**, 231, **232**, 236, *236*
Nagato class (Japan), **228**, 235, *241*
Naiad (UK), *142*
Nassau class (Germany), *157*, *167*
Natal (UK), *23*, 58–60, 81
Nauticus (German publication), 28
Naval Construction Department (UK), 14
Naval Ordnance Department (UK), 12
Navigatori class (Italy), *266*
Nedinsco, 132, 169
Nelson, Vice Admiral Horatio, 84
Nelson (UK), **112**, 115, **118**, 119, **140–1**, *143*
Nelson class (UK), *113*, 115, 119, *122*, 123, 126, *136*, 139
Neptune (UK), 78–9, *89*, 129, *281*, *285*
Nevada (USA), *37*, *57*, *98*, *192*, *205*, *217*, **282**, *290*
Nevada class (USA), 99, 180, *196*, *205*, *288*
New Jersey (USA), *215*
New Mexico (USA), **184**
New Mexico class (USA), *11*, *57*, *64*, *65*, *178*, *183*, *188*, *190*, *192*, *195*, *200*, **202**, *203*, *205*
New Orleans (USA), *225*
New Orleans class (USA), *64*, *205*, *225*, *226*
New York (USA), *102*, 178, 179, 182, *192*, *192*, **195**
New York class (USA), 187, *196*, *205*, *290*
Newcastle class (UK), *113*
Newfoundland (UK), *131*
Newport News class (USA), *284*
night action, 106, 113, 114, 143–5, 151, 153, 154–5, 205, 208–11, 213–16, 219–23, 230, 233, 237–8, 240, 267
night-vision binoculars, *237*
Nippon Optical Manufacturing Company, 237
Normandie class (France), *287*
Normandy landings, 155
Norfolk (UK), *143*, **146**, 148, 149
North Carolina (USA), **211**, **213**, **288**
North Carolina class (USA), *64*, 115, *203*, *210*, *212*, *213*, *235*
Northampton (USA), **201**, 216
Northampton class (USA), *64*, 190, *193*, *194*, 199, *199*, *200*, *237*
Nuremburg (Germany), 169

O

Ocean (France), *243*
Officine Galileo, 262, 263
Ogilvie, Captain F C A, 58–9, 60
Oklahoma (USA), 187, **196**, 204
Oktyabrskaya Revolutsiya (USSR), *268–70*, 278
Omaha class (USA), 65, 191, *191*, *194*, 284
Ontario (UK), *131*
Orion (UK), 20, **36**, 60, 79, **82–3**, 129
Orion class (UK), *32*, *36*, 60–1, 79
Ostfriesland (USA), *283*
Ostro (Italy), 263
Oyodo class (Japan), *241*

P

Packenham, Captain W C, 28, 36, 37, 38, 68
Panteleimon (Russia), 272, 273
Paris (France), **242–3**, *245*, 252
Paris class (France), *243*, *247*, 248
Parizhkaya Kommuna (USSR), *269*, 278
Pearl Harbor, *215*, 216, 235
Peirse, Rear Admiral, 78
pendulums, 19–20, 133
Pennsylvania (USA), **26–7**, *64*, **180–1**, *192*, 204, 221, 223
Pennsylvania class (USA), 178, *196*
Pensacola (USA), **194**, **226**
Pensacola class (USA), *122*, *184*, 191, *192*, *193*, 199
Perepelkin, Colonel Ia A, 274
Perth (Australia), 153
Perth class (UK), *113*
Pervenets (Russia), 270
Petr Velikiy (Russia), 273, 275
Philip (USA), *188*
Phoenix (USA), 224
Pillsbury (USA), *241*
plotting, 42–5, 93–4, *94*, 108, 113, 114–15, 121–4, 130–1, 144, 148, 169, 177–9, *178*, *179*, 182, 186, 191–2, 203, 210–11, **211**, 215, **229**, *234*, 245–6, 249, 250–1, 253–4, 256, 260–1, 263–4
  see also Combat Information Center
Plunkett, Lieutenant Reginald, *see* Drax, Admiral Lord
pointers, 19
Pollen, Arthur, 42–3, 44–5, 53–6, 58, 59–60, 61, 62, 63–5, *137*, 183, 274
Pollen system, 11, 28, 42, 44, 46, 53–61, *55*, 72, 91, 123, 131, *135*, 182, 184, 248, 274, 278
  see also Vickers-Pollen system
Poltava (Russia), *271*
Poore, Major, 23
Pope (USA), 241
Port Arthur, 37, 272
Porter class (USA), *226*
Portland (USA), *199*, 216, **224**
Portland class (USA), *200*
Portsmouth Naval Yard, 35
position-keeping
  analytic, 41, 44
  synthetic, 41–2
  see also Dumaresques; Dreyer Table; Vickers Clock
Poste à Calcul (PC), 248
Pothuau (France), 244, 246, 248, 253, 254
Precision Moderne, 251, 253
Prince Eugene (UK), *110*
Prince George (UK), 76, **100–1**
Prince of Wales (UK), *14*, 46, *137*, 142, 144, 148, 149–50, *174*
Princess Royal (UK), *34*, 35, 108, **108**
Principe Alfonso (Spain), *134*
Pringle, Captain, 23
Prinz Eugen (Germany), *39*, 169, 171–2, *172*
Progress in Gunnery (publication), 80, 143–4, 150
Provence (France), 154
Pugliesi, General, *264*
Punto Stilo, Battle of, 267

Q

Queen Elizabeth (UK), 52–3, *102*, *120–1*
Queen Elizabeth class (UK), 24, *32*, 115, *117*, *118*, 119, 130, 147, *151*, *228*
Queen Mary (UK), *32*, 60–1, 80, 161, 163
Queen Mary class (UK), 24

R

R class (UK), 25, *102*, 115, 117, 118
radar, 11
  electronic identification, 213
  fire-control, 28, 142, 147–8, *150*, *170*, *172*, *194*, *209*
  Japanese, *241*
night action, 143–5, 151, 211, 213–15
radio interference, 143
rangefinding, 114, 142, 146, 155, 174, 219–21, *219*, *220*, *221*, 223
splashes, *208*
supply, 141
radio, 114, 139, 143, 186, 223, 236, 258, 273, 279
Raimondo Montecuccoli (Italy), **264**
Raimondo Montecuccoli class (Italy), *266*
Raleigh (USA), *194*
Raleigh (Frobisher) class (UK), 46, *52*
Ramillies (UK), 48, 49, *51*, **103**, *118*
range-corrector, 42, 110
rangefinders
  22 ft, 147
  30 ft, 147
  anti-torpedo, *257*
  baffles, 23, *25*
  Barr & Stroud 5ft (FA Mk 1), 23
  Barr & Stroud 9ft, *23*, 25, 110, 244
  Barr & Stroud 15 ft, 24–5, 130, 178, 244, 260
  Barr & Stroud 18 ft, 274
  Barr & Stroud triplex, *38*
  control, 88–91, 93–4, 96–7, 108
  'dual-purpose', 265, **266**
  EU/SV-Anzeiger, 160, 161
  Fiske, 270
  FQ2 9ft, 23–4
  Liuzol'-Miakishev, 270
  long-base, *24*
  Mallock 8ft, 23
  multiple, 47–8, 89
  operators' station, *37*
  Pollen-Cooke, 274
  Ponthus-Therrode, 244
  quadruplex, *237*
  range projector, 176
  rate projector, *31*
  splinter shields, *27*
  stereoscopic, 23, *25*, 25–7, *39*, 153, 160, 179, 190, *192*, 203, *243*, 244, *247*, *249*, *253*, 264, 265, *267*, *276*, 278, 279
  in turrets, 24, *26*, *39*, 148, *149*, 178, *281*
  Watkin, 23
  see also Argo Clock; Dumaresques; gyro-stabilised rangefinders; sextants
rangefinding, 22–8
  bearing rate, 43–5, *43*, 158
  bracketing, 69–70
  dead time, 35, 245, 256
  geometric range, 27, 28
  gun flashes, 110
  gun range, 27
  ladder ranging, 104
  range rate, 27–31, *29*, *30*, *31*, 35, 38, 42–5, 48, 54, 88, 90, 91, 104, 121, 133, 154, 160, 163, 165, 169, 173–4, 176, 182–3, 205, 228, 229, 250, 260, 264, 272, 274, 277
  squadron, 236
  see also radar; radio; salvo fire; spotting
range-keeping, 12, *43*, *63*
  Ford Range-keeper, *31*, 42, *57*, *58*, *59*, 65, 104, *121*, 123, 134, *177*, 182–6, *187*, *192*, 250, 251, 252, 260
  German, 169–70
range telegraphs, *see* data transmission
rate of fire, 35, 38, 74, 78, 79, 99, 158
Renown (UK), 52, 117, 150
Renown class (UK), *102*
Repulse (UK), 111, 117, 146–7
Repulse class (UK), 48

INDEX

Resolution, 118, *118*, 147
Retvizan (Russia), 37, 272
Revenge (UK), 117
Riachuelo (Brazil, unbuilt), 65
Richelieu (France), **253**, **287**
Richelieu class (France), 244, *255*, 256
Richmond (USA), *191*, *194*, 225
Rivadavia class (Argentina), 186
River Plate, Battle of, *39*, 150, 155
ROCORD (Rate of Change of Range and Deflector indicator), 61, 131
Rodney (UK), 115, *118*, 119, 148–9, 155
Roma (Italy), **258**, 266
Roosevelt, Franklin D, 184
Roosevelt, Theodore, 176
Rossia (Russia), 272
Royal Navy Inspector of Target Practice, 20
Royal Oak (UK), *45*, *50*, 117, *118*
Royal Sovereign (UK), *118*, **145**, 150
Royalist (UK), 53
Rurik (Russia), *271*
Russo-Japanese War, 11, 22, 28, 36–9, 68, 69, *82*, 84, 233, 270, *271*, 272, 274, 288
Russo-Turkish War, 270

S

St Louis (USA), *202*
St Louis class (USA), *64*
St Vincent class (UK), *77*
Sakawa (Japan), **241**
Salt Lake City (USA), **194**, 225–6, **226**, 241
salvo firing, 35, 38
  air targets, 255–6
  brackets, 69–70, 94, 150, 161, 164, 176, 204, *208*, 246, 273
  broadsides, 136, 138, *180*, *183*, 203, 236
  concentration fire, 81, 122, 139, 186, 236, *257*, 258
  deflection group, 138, 150
  directors, 73–81
  dispersion, 203–4, 246, 257, 266
  ladders, 101, 103–4, 105, 108, 135, 136–9, 142, 161, 163, 205, 209, 219, 228, 236, 246, 256, 258, 267, 277
  rapid group, 94, 138–9, 148, 163, 166, 272
  rocking, 205, 219, 220
  straddles, 69, 103, 136–8, 142, 148, 161–3, 174, 204, *216*, 223, 224, 225–6, 236, 246, 258
  trend groups, 136
  zigzag group, 136–9, 142, 144, 150, 151, 163, 205
Samar, Battle of, 69, 223, 232, 235, 241
San Francisco (USA), 209, 211, 213, 216
San Francisco class (USA), 200
Sao Paulo (Brazil), 186
Satsuma (Japan), 233
Savage (UK), 145, 146
Savannah (USA), 285
Savo Island, Battle of, 240, 241
Scapa Flow, *161*, 186, 203, 204, 285
Scarborough Raid, 107–8
scartometry, *see* splashes
Scharnhorst (Germany), 111, 142, 144, 145, 150, 154
Scharnhorst class (Germany), 169
Scheer, Vice Admiral Reinhard, *94*
Schlesien (Germany), 167
Scott, Admiral Sir Percy, 18–20, *70*, 72, 73–5, 76, 78, 79, 80, 158, 176, 179, 245, 272, 282
Scylla (UK), 18
searchlights, 153, *195*, 215, 216, 220, 230, 238
Selbourne, Lord, 21

Serrano class (Chile), 168
sextants, 22, 28, 158, 160, 270
Seydlitz (Germany), 108, 109, 111, *157*, 159, *159*, 163
Shannon, 81
Sheffield (UK), *131*, 155
shells
  armour-piercing (AP), 55, 110, 115, 154, 222–3, 225, 285–9, **290**, **291**
  armour-piercing, capped (APC), 233–5, **288**, 289
  capped piercing common (CPC), 233, 289
  coloured splashes, 155, 234, 258
  flashless powder, 153, 155
  French, *245*, 257
  high-capacity (HC), 222, 225, 289, 290
  high-explosive (HE), 154, 288, 289
  high-velocity, 115, 158–9, *161*, 282, 291
  Italian, *263*
  Jutland, 111
  'K' device, 155, 258
  long, 146–7
  lyddite, 109–10, 154, 233, 289
  propellant, 282–5
  Russian, *271*
  shellite, 289
  starshell, 145, 155, 209, 220, 230, 237, 238, 241
supercharges, 147
super-heavy, *215*, **290**
Type 91 (Japan), 236, 241
Shepherd, Sir Victor, 98
Shinyo (Japan), 2233
ship motion, 18–21, 42, 44, 45, 74–5, 76, 105, 145–6, 229
shore bombardment, 110, *137*, 153, 155, 220, 225, 289
Siemens (UK), 33
Siemens-Halske (Germany), 132, 164, 167, 263
Sims, Lieutenant William S, 20, 176
Sino-Japanese War, 176
Slava (Russia), 278
smoke protection, *12*, 119, 220, 226, 236
sokutekiban (target inclination device), 231, **231**, **233**, **234**, **235**, **239**, **240**
Solomon Islands, Battle of, 210, 225, 240
Somers class (USA), **226**
South Carolina (USA), *178*, 179
South Carolina class (USA), 203
South Dakota (USA), 154, *208*, **212**, 220, 241
South Dakota class (USA), 205, *210*, *212*, 235
Southampton class (UK), *113*
Sovietskii Soyuz class (USSR), 279
Spanish-American War, 176
Spee, Admiral Maximilian Graf von, 106
Sperry, Elmer Ambrose, 184
Sperry Gyroscope, *62*, 182, 278
  see also fire-control system
splashes
  chasing, 241
  coloured, 155, 234, 258
  concentrated fire, 81, 105
  'deviation' method, 275
  director control, 104
  distinguishing, 105, 108, 109, 110, 148, 155, 163, 219, 273
  early British experiments, 68
  full salvoes, 180
  mean point of impact, 142
  radar, *208*
  scartometry, 244, *247*, 264–5, 266, 266, 278

slicks, 176
starshell, 209
spotting, 35, 68–71, 74, 76, 78, 80, 89, 93, 94, 96, 107, 108–9, 111, 119–21, 135–9, 148, 150, 155, 160, 176, 180, 190, *193*, 204, 205, 219, 221, 245, 246, 254, 264–5, 273, 277, 278
  see also air-observation
stadimeter, 22
Stalingrad (USSR), 275
Stalingrad class (USSR), **277**, 279
Strasbourg (France), 249
Sturdee, Admiral Doveton, 85
submarines, 209, 239, 262
Sumatra (Holland), 168, 169
Superb (UK), *33*, *131*, 147, 288
Surigao Strait, Battle of, 11, *216*, *218*, *219*, 221–4, 241, *290*
Sverdlov class (USSR), *273*, 275, *275*
Sverige (Sweden), **167**, 168
Sverige class (Sweden), 167
Svobodnaya Rossiya (Russia), 277
Swiftsure (UK), *131*, 147
Sydney (Australia), *29*, 151

T

Takao class (Japan), *239*
Taranto, Battle of, *265*
Tassafaronga, Battle of, 216
telephone exchange, *see* transmitting station
Temeraire (UK), *104*
Tennessee (USA), *28*, *178*, **185**, *189*, 191, 221, 223, 224
Tennyson d'Eyncourt, Sir Eustace, 99
Terrible (UK), 20
Texas (USA), *105*, 178, 179, 182, 192, *192*, 195
Texas class (USA), *175*
Thomas (USA), *188*
Thomas Cooke, 58
Thomsen, Admiral von, 158
Thunderer (UK), *24*, 61, 79, 80, 179
Thursfield, Captain H G, 92–3
Tiger (UK), *46*, 80, *92*, 109
Tiger class (UK), 147
Tirpitz, Admiral Alfred von, 85, 158
Tirpitz (Germany), 10, 148, **173**
Togo, Admiral, 37–8, 96
Tone class (Japan), *239*
torpedo craft, 22
torpedoes
  American, 208, 213
  British, 267
  on battleships, *116*, 244
  bombers, 119
  bulging, 114, *118*
  destroyer-launched, 101
  Japanese, 209–10, 215, 216, 220, 236–7, 238, 239, 241
  Jutland, 96
  rangefinders, *257*
  tactics, 85–6, 88, 90, 91, *92*, 113–14
  underwater armour, 235
  Whitehead, 56
  see also gyro-controlled torpedoes
Tosa (Japan), 233
Toulon naval-gunnery school, 246
'Town' class (UK), *113*, 129–30
transmitting station, 36, 42, 46, 62, 78, 119–21, 129–30, 144–5, 146, 148, 150, 176–7, 230–2
  see also Poste à Calcul
transmission systems, 125–9
Tre Kronor class (Holland), 284
'Treaty' class (USA), 180
Trento (Italy), 263
Trento class (Italy), *263*, 264
Trieste (Italy), 263
Triumph (UK), 288

Tsarevich (Russia), 37
Tsushima, Battle of, 36, 38, 84, 109, 235, 272–3
Tudor, Rear Admiral Frederick, 99

U

Uganda (UK), **149**
Usborne, Admiral Captain C V, 98, 113–14
Utah (USA), 179, 182

V

Valiant (UK), 115, 155
Vanguard (UK), *43*, 130, 147, **151**, **153**, 285
Vauquelin class (France), *257*
Venerable (UK), 33
Vengeance (UK), 74
Vermont (USA), 180
Versailles Treaty, 167
Vickers, 31–3, 57, 61, 63, 74, 80, 133–4, **134**, 179, 188, 190, 228, 251, 252, *271*, 278, 287
  see also Admiralty Fire-Control Clock
Vickers Clock, 41–2, *41*, 46, 48, 56
Vickers-Pollen system, 133, **134**
Victoria (UK), 84
Victorious (UK), *17*, 33
Vietnam War, *215*
Vincennes (USA), *64*, 200, *204*
Virginia (USA), 177
Virginia class (USA), 179
Viribus Unitis (Austria-Hungary), **165**, *271*
Viribus Unitis class (Austria-Hungary), 165
Vladivostok, 272
Volya (Russia), 274
Von der Tann (Germany), 161

W

Warrior (UK), 166
Warspite (UK), *102*, *113*, 115, 117, 119, **136**, 139, 150
Washington (USA), **206–7**, **208**, 219–21
Washington Treaty, 113, *113*, 115, *127*, 180, *195*, 265, **287**, 290
West Virginia (USA), *64*, 187, 188, **192**, 193, **216**, 221, 223
Westfalen (Germany), **156–7**
Wichita (USA), *64*, **204**
Wiesbaden (Germany), 166
Willem Van Der Zaan (Holland), 168
Wilmington class (USA), 224
Wilson, Admiral Sir Arthur Knyvet, 56
Wisconsin (USA), **214**, 215
Worcester class (USA), *64*, 224–5
Wyoming (USA), *27*, *28*, *178*

Y

Yakovlev, Lieutenant, 279
Yamagumo (Japan), 222
Yamashiro (Japan), 223, **234**
Yamato (Japan), 10, **15**, 231, **237**
Yamato class (Japan), *215*, 236, **290**, **291**
Yarra (UK), 155
Yellow Sea, Battle of, 36, 39, 68, 96
York (UK), **129**
York class (UK), *113*

Z

Zara class (Italy), 263
Zeiss, 25, 132, 167, 168, 179, **269**, 274
zentralnii artilleriiskii post (ZAP), 273–4, 277